Pre-Algebra & Word Problem Applications

ANNA FISHER
Bloomfield College

KENDALL/HUNT PUBLISHING COMPANY
4050 Westmark Drive   Dubuque, Iowa 52002

*Cover Illustration by Vladimir Guillaume

Copyright © 2005 by Anna Fisher

ISBN 978-0-7575-4515-3

Kendall/Hunt Publishing Company has the exclusive rights to reproduce this work, to prepare derivative works from this work, to publicly distribute this work, to publicly perform this work and to publicly display this work.

All rights reserved. No part of this publication may be reproduced, stored in a retrieval system, or transmitted, in any form or by any means, electronic, mechanical, photocopying, recording, or otherwise, without the prior written permission of Kendall/Hunt Publishing Company.

Printed in the United States of America
10 9 8 7 6 5 4

CONTENTS

CHAPTER 1 — 1

    1.1 Real Numbers 1

    1.2 Reading Whole Numbers 5

    1.3 Rounding Off Numbers 9

    1.4 Primes, Divisibility, and Factor-Trees 13

    1.5 Computations with Whole Numbers 19

    1.6 Exponents and Order of Operations with Whole Numbers 23

    1.7 Application with Whole Numbers 29

CHAPTER 2 — 41

    2.1 Addition and Subtraction with Signed Numbers 41

    2.2 Multiplying and Dividing Signed Numbers 49

    2.3 Exponents and Order of Operations with Integers 53

    2.4 From Words to Algebra 57

CHAPTER 3 — 61

    3.1 Fractions and Mixed Numbers 61

    3.2 Equivalent Fractions 69

    3.3 Multiply and Divide with Fractions and Mixed Numbers 77

    3.4 Add and Subtract with Fractions and Mixed Numbers 85

    3.5 Exponents and Order of Operations with Fractions 95

    3.6 Applications with Fractions 99

    3.7 Exponents and Order of Operations with Signed Fractions 113

    3.8 More Words to Algebra 117

CHAPTER 4 — 121

    4.1 Reading and Writing Decimals 121

    4.2 Rounding Off Decimals 125

4.3 Add, Subtract, Multiply, Divide with Decimals   129

4.4 Converting Fractions and Decimals   137

4.5 Applications with Decimals   141

4.6 Order of Operations with Fractions, Decimals, and Signed Numbers   147

CHAPTER 5   151

5.1 Ratios   151

5.2 Proportions   157

5.3 Applications Involving Proportions   163

CHAPTER 6   169

6.1 Percents   169

6.2 Converting Decimals and Fractions to Percents   173

6.3 Percentage Problems   177

6.4 Applications Involving Percents   183

6.5 Applications Involving Simple Interest   191

CHAPTER 7   197

7.1 Variables and Formulas   197

7.2 Properties of Algebra   203

7.3 Simplify Algebraic Expressions   209

7.4 Solving Equations Using the Addition Property and Multiplication Property   215

7.5 More on Solving Equations   223

7.6 More Words to Algebra and Solve   229

CHAPTER 8   233

8.1 An Introduction to Sets   233

8.2 Operations on Sets   239

8.3 Equivalent Sets and Cardinal Numbers   245

8.4 Venn Diagrams and Applications   249

CHAPTER 9   253

9.1 Outcomes   253

9.2 The Fundamental Principle of Counting   259

9.3 Probability of Events   267

1.1 REAL NUMBERS

CHAPTER 1

The number system is one of the foundations of mathematics. It is important to know how the numbers are categorized and named so that there is a common language when discussing numbers.

Natural or Counting Numbers: the set of $\{1, 2, 3, \ldots\}$
Whole Numbers: the set of counting numbers including zero, $\{0, 1, 2, 3, \ldots\}$
Integers: the set of positive and negative whole numbers, $\{\ldots, -3, -2, -1; 0, 1, 2, 3, \ldots\}$

RATIONAL NUMBERS

There are two versions of the definitions:

1) the set of $\{\frac{a}{b} \mid a \text{ and } b \text{ are integers and } b \neq 0\}$

   Examples: $\frac{3}{5}$ numerator and denominator are both integers

   -8 numerator and denominator are both integers, *any integer is understood to be over one*

   0.5 can be rewritten as $\frac{5}{10}$ numerator and denominator are both integers

2) any decimal that repeats or any decimal that terminates.

   Examples: 4 a terminated decimal

   $1.333\ldots$ a repeating decimal

   $5\frac{1}{4}$ can be rewritten as 5.25 a terminated decimal

   $\sqrt{4}$ can be simplified as 2   a terminated decimal

Irrational Numbers: any decimal that does not repeat and does not terminate

   Examples: $\pi = 3.1415927\ldots$ non-repeating and non-terminated decimal

   $\sqrt{3} = 1.7320508\ldots$ non-repeating and non-terminated decimal

   $\frac{\sqrt{5}}{3} = 0.745355993\ldots$ non-repeating and non-terminated decimal also, numerator is not an integer

---

1.) Pages 009–018, 037–042, 045–047, 052–059, 062–077, 080–087, 115–127, 129–132, 141–143, 147–152, 184–187, 199–203, 206–209, from *Math for College Students: Arithmetic with Introductions to Algebra and Geometry*, 5th Edition by Ronal Straszkow. © 1997 by Kendall/Hunt Publishing Company. Used with permission.

Real Numbers: the set of rational numbers and irrational numbers

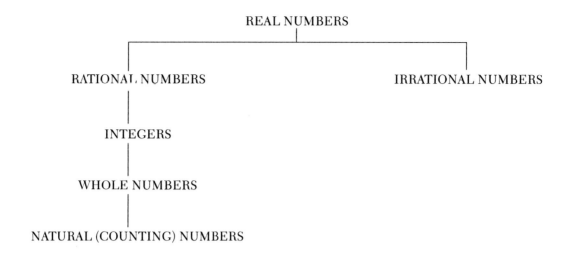

PRACTICE PROBLEMS

1) State all the terms associated with each number.

    a) -25 *solution:* real number, integer, rational number

    b) 11.8 *solution:* real number, rational number

    c) $-\sqrt{5}$ *solution:* real number, irrational number

    d) 200 *solution:* real number, natural (counting) number, whole number, integer, rational number

2) State whether each of the following statements is true or false.

    a) $5.326952\ldots$ is an irrational number.   *solution:* True; the decimal is non-repeating and non-terminated

    b) $\frac{2}{3}$ is not a real number.   *solution:* False; $\frac{2}{3}$ is a rational number and therefore, a real number

    c) An integer is also a rational number.   *solution:* True; since an integer has an understood denominator of one, then both numerator and denominator are integers; or an integer is a terminated decimal

EXERCISE 1.1 SET A

A. State whether each of the following statements is true or false.

1) Some irrational numbers are not real numbers.

2) $\dfrac{4}{9}$ is an irrational number.

3) 0 is not an integer.

4) An integer is also an irrational number.

5) A rational number is either an integer or a non-integer.

6) $\dfrac{8}{9}$ is not a whole number but it is a rational number.

7) 5.326952... is an irrational number.

8) $\sqrt{5}$ is not a real number.

9) 0.3232... is a rational number.

10) -3.25 is a rational number.

B. Given the set of numbers

$$A = \{-5,\ 8,\ 2.7,\ -\dfrac{4}{5},\ \sqrt{7},\ -\dfrac{2}{5},\ 0.2835843...,\ 0,\ 0.25,\ 55,\ 2.58,\ -\sqrt{3},\ \pi\}$$

List the numbers in the set that are:

11) Integers

12) Whole numbers

13) Rational numbers

14) Irrational numbers

15) Real numbers

EXERCISE 1.1 SET B

A. State whether each of the following statements is true or false.

1) All irrational numbers are real numbers.

2) -2000 is an integer and a rational number.

3) A whole number is also an irrational number.

4) $\dfrac{2}{3}$ is a rational number as well as an irrational number.

5) 0.8 is a rational number.

6) $-4.525252\ldots$ is an integer.

7) $\sqrt{36}$ is a whole number.

8) 95 is a natural number, whole number, integer, rational number, and a real number.

9) An irrational number is also an integer.

10) $\dfrac{-5}{\sqrt{3}}$ is a rational number.

B. Given the set of numbers

$$A = \{-5.3434\ldots, \ -78, \ \tfrac{1}{3}, \ 6.50, \ 0, \ \sqrt{8}, \ -0.11, \ \sqrt{81}, \ \pi, \ 3\tfrac{4}{5}, \ \tfrac{\sqrt{6}}{5}\}$$

List the numbers in the set that are:

11) Whole numbers

12) Integers

13) Rational numbers

14) Irrational numbers

15) Real numbers

1.2 READING WHOLE NUMBERS

All whole numbers can be formed using the **digits** 0, 1, 2, 3, 4, 5, 6, 7, 8, and 9. This can be done since a digit's position in a number gives it a specific meaning called its **place value.**

For example:

In 576,

> the 5 means 5 hundreds (500)
> the 7 means 7 tens (70)
> the 6 means 6 ones (6).

Thus, 576 is read "five hundred seventy-six."

In 4,208,

> the 4 means 4 thousands (4000)
> the 2 means 2 hundreds (200)
> the 0 means 0 tens
> the 8 means 8 ones (8).

Thus, 4,208 is read "four thousand, two hundred, eight."

Number:
The dentist numbed her with novocaine.

The following is a chart for the place values in our number system:

$$\underbrace{6\ 7\ 8}_{\text{billions}},\underbrace{3\ 1\ 0}_{\text{millions}},\underbrace{4\ 5\ 6}_{\text{thousands}},1\ 2\ 7$$

where the digits are labeled (left to right): hundreds-billions, ten-billions, billions, hundreds-millions, ten-millions, millions, hundreds-thousands, ten-thousands, thousands, hundreds, tens, ones.

In the United States, large numbers are separated into groups of three digits and read according to this place value chart.

For example: In the number 17832536 you would start from the right hand side of the number and mark off every three digits with a comma as follows:

17,832,536

You then read each group of digits that was marked off, followed by the value for that group as noted on the place value chart.

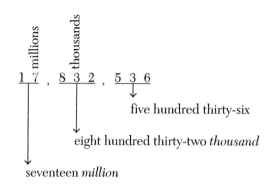

Thus, the number is read "seventeen million, eight hundred thirty-two thousand, five hundred thirty-six."

REMEMBER: When placing commas to separate the digits, always start from the right hand side of the number and work your way to the left, marking off every three digits.

NOTE: The word "and" is not used when writing the word equivalent for whole numbers. It is used to indicate the placement of a decimal point, as we shall see later. For example, 5008 is read "five thousand, eight." It is *not* read "five thousand and eight."

NOTE: Compound numbers from 21 to 99 are written with a hyphen, such as twenty-one or ninety-nine.

PROBLEMS

1) Write 10,487 in words.

    Answer: ten thousand, four hundred eighty-seven

2) Write 2,003,500 in words.

    Answer: two million, three thousand, five hundred

3) In the number 325,467 what does the:

    Answers:

    5 represent?   5 thousands (5000)

    4 represent?   4 hundreds (400)

    2 represent?   2 ten thousands (20,000)

4) Write using numbers:

    Answers:

    a) four hundred twenty-two million, ten   a) 422,000,010

    b) thirty-five billion, twenty-seven million, two thousand, six   b) 35,027,002,006

Name _____   Date _____

EXERCISE 1.2

In problems 1–7, write each number in words.

1) 852

2) 4,256

3) 17,109

4) 3,057,010

5) 14,100,700

6) 946,003

7) 1,357,926,183

1) _____

2) _____

3) _____

4) _____

5) _____

6) _____

7) _____

In problems 8–16, write each expression using numbers and commas.

8) seven hundred forty-five

9) fifty thousand, sixty-eight

10) one hundred five thousand, six

11) forty million, thirty-six

12) five million, seven thousand, two hundred thirty-eight

13) twelve billion, fifteen million

14) eighty-nine billion, eighty-nine

15) three hundred thirteen million, seven hundred ten thousand

16) seven hundred twelve million, four hundred twenty-two

8) _____

9) _____

10) _____

11) _____

12) _____

13) _____

14) _____

15) _____

16) _____

In problems 17–21, consider the number 5,876,492.

17) What does the 5 represent?

18) What does the 9 represent?

19) What does the 8 represent?

20) What does the 6 represent?

21) What does the 7 represent?

17) _____

18) _____

19) _____

20) _____

21) _____

1.3 ROUNDING OFF NUMBERS

If the number of people living in your city is 178,952, the road sign would probably read as in the illustration on the left.

The sign gives the approximate population of the city. It is just an estimate of the number of people living there. We say that the population has been rounded off. When you round off a number, it is close to the original number. Rounded numbers are easier to visualize and simpler to work with than the original numbers. Rounded numbers can also be used to estimate answers in complicated math problems.

Whole Numbers:
The digits in that sign have holes in them. They must be hole-numbers.

ROUNDING OFF TO THE NEAREST TEN

If you were to count by tens you would say 10, 20, 30, 40, 50, 60, 70, etc. These are called the multiples of ten.

To round off a given number to the nearest ten, you must find the multiple of ten that is closest to the number.

For example: To round off 37 to the nearest ten, the answer would be 40, since 37 is between 30 and 40 but is closer to 40. We say 37 rounds off to 40.

On the other hand, 32 rounds off to 30, since it is closer to 30. We say 32 rounds off to 30.

Now 35 is halfway between 30 and 40. In this case, round it off to the higher multiple of ten. Thus, 35 rounds off to 40.

PROBLEMS

1) Round off 86 to the nearest ten. *Answer:* 90
   86 is between 80 and 90 but is closer to 90.

2) Round off 723 to the nearest ten. *Answer:* 720
   723 is between 720 and 730 but is closer to 720.

3) Round off 2695 to the nearest ten. *Answer:* 2700
   Since 2695 is halfway between 2690 and 2700, round off to the higher multiple, 2700.

ROUNDING OFF TO THE NEAREST HUNDRED

The multiples of a hundred are 100, 200, 300, 400, 500, 600, and so on.

To round off a given number to the nearest hundred, you must find the multiple of a hundred that is closest to the number.

For example: 768 rounds off to 800 since 768 is between 700 and 800 but is closer to the higher multiple of a hundred, 800.

725 rounds off to 700 since it is closer to the lower multiple of a hundred, 700.

ROUNDING OFF TO OTHER PLACES

You round off to other places in our number system using the same procedure as was used in rounding off to the nearest ten or hundred. You round off to the lower multiple if a given number is less than half the way between two multiples. You round off to the higher multiple if it isn't. An easy way to decide this is as follows:

> **BASIC ROUND OFF RULE**
> Look at the digit to the right of the place you are rounding to. If that digit is 5 or more, round off to the higher multiple. Otherwise, round off to the lower multiple.

Example 1

Round off 28,752 to the nearest thousand.

$$28{,}752 \approx 29{,}000 \quad (\approx \text{ means "is aproximately equal to."})$$

thousands place ↑ — digit to the right is more than 5

Thus, you round off to the higher multiple of a thousand by changing the 8 to a 9, getting 29 thousands (29,000).

Example 2

Round off 413,207,593 to the nearest million.

$$413{,}207{,}593{,} \approx 413{,}000{,}000$$

millions place ↑ — digit to the right is less than 5

Thus, you round off to the lower multiple of a million by leaving the 3 unchanged, getting 413 millions (413,000,000). *Notice:* All the digits to the left of the millions place are unchanged while those to the right are replaced with zeros.

PROBLEMS

4) Round off 865,000 to the nearest ten thousand.   *Answer:* 870,000

Since the digit to the right of the ten thousands place is a 5, round off to the higher multiple by adding one to the ten thousands place.

5) Round off 96,379 to the nearest thousand.   *Answer:* 96,000

Since the digit to the right of the thousands place is less than 5, do not increase the 6 in the thousands place. Leave the first two digits as 96 and replace the 379 with zeros.

Name _____   Date _____

EXERCISE 1.3

In problems 1–7, round off each number to the nearest ten.

1) 53

2) 76

3) 25

4) 506

5) 1473

6) 5895

7) 6997

1) _____

2) _____

3) _____

4) _____

5) _____

6) _____

7) _____

In problems 8–14, round off each number to the nearest hundred.

8) 549

9) 872

10) 650

11) 14,736

12) 27,864

13) 179,950

14) 1,538,276

8) _____

9) _____

10) _____

11) _____

12) _____

13) _____

14) _____

In problems 15–22, round off 215,749,538 to the place indicated.

15) tens place

16) hundreds place

17) thousands place

18) ten thousands place

19) hundred thousands place

20) millions place

21) ten millions place

22) hundred millions place

15) _____

16) _____

17) _____

18) _____

19) _____

20) _____

21) _____

22) _____

1.4 PRIMES, DIVISIBILITY, AND FACTOR-TREES

PRIMES

A **prime number** is a whole number greater than one that is evenly divisible by only 1 and itself. If a number is evenly divisible by other numbers besides 1 and itself, it is a composite number.

2, 3, 5, 7, 11, 13, 17, 19, 23, 29, . . . are prime numbers.

4, 6, 8, 9, 10, 12, 14, 15, 16, 18, . . . are composite numbers.

It has been proved that there are an infinite number of primes and that every composite number can be represented as a product of primes. The Greek mathematician Eratosthenes (275–195 B.C.) created an interesting mechanism for finding prime numbers, the Sieve of Eratosthenes. For example, to find the prime numbers from 2 to 200, write down all the whole numbers from 2 to 200. The first number, 2, is a prime. Draw a box around it and cross out all the other multiples of 2 (every second number: 4, 6, 8, 10, 12, . . .). The next number not crossed out, 3, is a prime. Draw a box around it and cross out all the other multiples of 3 (every third number: 6, 9, 12, 15, 18, 21, . . .). Continuing in this manner you will find that there are 46 prime numbers less than 200. If you wrote down more whole numbers, you could use this technique to find more of the prime numbers.

2	3	4	5	6	7	8	9	10	11	12	13	14	15	16	17	18	19	20	21
22	23	24	25	26	27	28	29	30	31	32	33	34	35	36	37	38	39	40	41
42	43	44	45	46	47	48	49	50	51	52	53	54	55	56	57	58	59	60	61
62	63	64	65	66	67	68	69	70	71	72	73	74	75	76	77	78	79	80	81
82	83	84	85	86	87	88	89	90	91	92	93	94	95	96	97	98	99	100	101
102	103	104	105	106	107	108	109	110	111	112	113	114	115	116	117	118	119	120	121
122	123	124	125	126	127	128	129	130	131	132	133	134	135	136	137	138	139	140	141
142	143	144	145	146	147	148	149	150	151	152	153	154	155	156	157	158	159	160	161
162	163	164	165	166	167	168	169	170	171	172	173	174	175	176	177	178	179	180	181
182	183	184	185	186	187	188	189	190	191	192	193	194	195	196	197	198	199	200	

Example 1

Determine if the following numbers are prime or composite.

a. 31   Prime (It is only divisible by 1 and 31.)

b. 326 Composite (It is divisible by 2 and 163.)

c. 111 Composite (It is divisible by 3 and 37.)

DIVISIBILITY

In determining if a given number is prime or composite, you need to decide if a number besides 1 and itself divides evenly into the given number. Below are some facts about divisibility that will help you determine what numbers divide evenly into a given number.

Divisibility by 2: If a number ends in 0, 2, 4, 6, or 8, it is an even number. Even numbers are divisible by 2.

96 and 1278 are divisible by 2, since they are even numbers.

Divisibility by 3: If the sum of the digits of a number is divisible by 3, then the number is divisible by 3.

441 and 3417 are divisible by 3 since the sum of their digits are divisible by 3. In 441, 4 + 4 + 1 = 9 and 9 is divisible by 3. In 3417, 3 + 4 + 1 + 7 = 15 and 15 is divisible by 3.

Divisibility by 5: If a number ends in 5 or 0, it is divisible by 5.

935 and 2480 are divisible by 5 since they end in 5 and 0.

Divisibility by 10: If a number ends in 0, it is divisible by 10.

560 and 67,890 are divisible by 10 since they end in 0.

Divisibility by other numbers: There are no shortcuts to determine divisibility of most other numbers. You may simply have to determine if any of the prime numbers up to the square root of the number are divisors by actually performing the division.

The first prime that divides into 247 is 13 and the first prime that divides into 1141 is 7.

PROBLEMS

Is 510 divisible by:

1) 2?

2) 3?

3) 5?

4) 7?

Answers:

1) Yes, it is an even number.

2) Yes, sum of its digits is 6, which is divisible by 3.

3) Yes, it ends in 0.

4) No, dividing by 7 leaves a remainder of 6.

FACTOR-TREES

The divisibility facts can be helpful in writing a number as the product of other numbers. This process is called **factoring.** There may be different ways to factor a number. For example, the number 12 can be factored using whole numbers in the following ways, $12 = 1 \times 12$, $12 = 2 \times 6$, $12 = 3 \times 4$, or $12 = 2 \times 2 \times 3$. If all the factors are prime numbers, the number is expressed as the product of primes. Thus, $12 = 2 \times 2 \times 3 = 2^2 \times 3$ gives 12 factored into primes. We will spend the rest of this section examining how to factor numbers into primes realizing that it will be useful when operating with fractions.

A good technique for factoring a whole number into primes involves making a tree-like structure of the factors of the number. For example, to write the number 72 as a product of primes, find any two whole numbers that have a product of 72 and then express those numbers as a product of other whole numbers. If a number is prime, it is circled. If it is not, it is factored again. This process gives a diagram called a **factor-tree.** The product of all the encircled primes is the prime factorization of 72.

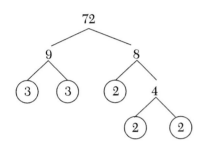

$9 \times 8 = 72$

$3 \times 3 = 9$ and $2 \times 4 = 8$

$2 \times 2 = 4$

Thus, $72 = 3 \times 3 \times 2 \times 2 \times 2 = 3^2 \times 2^3$

Notice that repeated factors are written with exponents.

In a factor-tree, you could start with any two whole numbers that have a product of 72. If you continue factoring until you arrive at prime numbers, you will get the same result.

Example 2

Factor 1000 into primes.

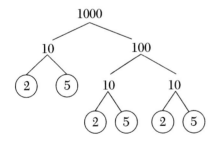

$10 \times 100 = 1000$

$2 \times 5 = 10$

$2 \times 5 = 10$

Thus, $1000 = 2 \times 5 \times 2 \times 5 \times 2 \times 5 = 2^3 \times 5^3$

Name _____   Date _____

EXERCISE 1.4

In problems 1–6, determine if each number is prime or composite.

1) 17

2) 23

3) 21

4) 49

5) 315

6) 441

In problems 7–24 factor each number into primes.

7) 18   8) 24

9) 46   10) 34

11) 42   12) 66

13) 5500   14) 3500

15) 75   16) 245

17) 1375   18) 1625

19) 660   20) 650

21) 222   22) 387

23) 133   24) 143

1) _____

2) _____

3) _____

4) _____

5) _____

6) _____

7) _____

8) _____

9) _____

10) _____

11) _____

12) _____

13) _____

14) _____

15) _____

16) _____

17) _____

18) _____

19) _____

20) _____

21) _____

22) _____

23) _____

24) _____

1.5 COMPUTATIONS WITH WHOLE NUMBERS

Operations on numbers refer to addition, subtraction, multiplication, and division.
For each operation, the parts to the problem are given terms.

Addition Problem:	$8 + 2 = 10$	8 and 2 are called the addends; 10 is called the sum
Subtraction Problem:	$8 - 2 = 6$	8 is called the minuend; 2 is called the subtrahend; 6 is called the difference
Multiplication Problem:	$8 \times 2 = 16$	8 and 2 are called the factors; 16 is called the product
Division Problem:	$8 \div 2 = 4$	8 is called the dividend; 2 is called the divisor; 4 is called the quotient.

NOTE:

1) The multiplication operation can be expressed using different symbols.

   Examples: "$\times$" represents the multiplication operation 3×5
   "$\cdot$" represents the multiplication operation $3 \cdot 5$
   "( )" parenthesis represents the multiplication operation $3(5)$ or $(3)(5)$

2) The division operation can be expressed using different symbols.

   Examples: "$\div$" represents the division operation $10 \div 5$
   "$-$" fraction bar represents the division operation $\dfrac{10}{5}$
   "$\overline{)}$" represents the division operation $5\overline{)10}$

NOTE:

In finding the answer for a division problem and using long division, the parts of the problem must be placed in the correct places.

$$a \div b = \frac{a}{b} = b\overline{)a}$$

Examples: $12 \div 4 = \dfrac{12}{4} = 4\overline{)12}$

$4 \div 12 = \dfrac{4}{12} = 12\overline{)4}$

EXERCISE 1.5

Find each answer.

1) 2095 + 1345 + 3456

2) 23458 + 1343 + 907

3) 25609 + 8888

4) 45678 + 35 + 1005 + 8 + 250

5) 5699 − 3144

6) 8000 − 1256

7) 10450 − 9004

8) 7421 − 475

9) 24 · 12

10) 308 · 45

11) 142 · 220

12) 303 · 101

13) 1000 · 560

14) 300 · 40

15) 1300 ÷ 4

16) 96450 ÷ 5

17) 2160 ÷ 12

18) 14322 ÷ 22

19) 96590 ÷ 10

20) 4560 ÷ 10

21) 78500 ÷ 100

22) 35,987,500 ÷ 100

23) 150,000 ÷ 1000

24) 633,500,000 ÷ 10,000

25) Find the sum of twenty-three, two hundred three, and two thousand three hundred three.

26) Find the difference of five thousand and six hundred forty-four.

27) Find the product of eighty-five and eleven.

28) Find the quotient of three thousand, four hundred seventy-one and three.

29) Find 3205 more than 23900.

30) Find twice the amount of 34,568.

31) Find 566 less than 1000.

32) Find 3,579,132 divided by 4.

1.6 EXPONENTS AND ORDER OF OPERATIONS WITH WHOLE NUMBERS

A multiplication problem in which a number is repeatedly multiplied by itself can be rewritten in shorthand form.

i.e. $4 \times 4 \times 4 \times 4 \times 4 \times 4$ can be rewritten as 4^6. 4^6 is in exponent form.

EXPONENT NOTATION

$a^b = a \cdot a \cdot a \cdot a \cdot a \ldots (b \ times)$

where a is the base (the number repeatedly multiplied)
b is the exponent or power (indicates how many times to repeatedly multiply the base)

Example: Find 5^3.
Solution: 5^3 indicates that the base, 5, will be multiplied 3 times.
Therefore, $5^3 = 5 \cdot 5 \cdot 5 = 125$.

DEFINITION

$a^1 = a$ any expression to the first power is the expression itself.

$a^0 = 1$ any expression to the zero power is one.

Examples: 1) $543^1 = 543$

2) $\left(\dfrac{2}{3}x\right)^0 = 1$

NOTE:

10^n, n is a whole number, can easily be computed by using the fact that there will be n number of zeros.

Examples: $10^1 = 10$
$10^2 = 100$
$10^3 = 1000$
$10^4 = 10,000$
.
.
.
$10^9 = 1,000,000,000$

When writing an essay, there is an order to follow. The essay should have an introduction, a body, and then a conclusion.

When finding an answer with several operations in math, there is an order to follow. This section will give you the order of operations.

For the expression 5 + 10 · 2 there are two possible answers.

If you computed 5 + 10 first, and then multiplied by 2, the answer is 100.

If you computed 10 · 2 first, and then added 2, the answer is 25.

Imagine if we allowed people to compute that expression in whichever order they wanted. There would be chaos. The world and its advancements as we know it today would probably not exist. That is why we have in place the order of operations.

ORDER OF OPERATIONS

1. Simplify any operations WITHIN grouping symbols such as parenthesis (), brackets [], braces {}, absolute value ||, and above or below fraction bar.

   Note: If the expression contains more than one grouping symbol, then work from the inner most grouping symbols to outer most grouping symbols.

2. Simplify any exponents/powers and radicals.
3. Simplify division and multiplication as they occur from left to right.
4. Simplify subtraction and addition as they occur from left to right.

Note: The order of operations still apply within the grouping symbols.

PRACTICE PROBLEMS

1) 5 + 10 · 2   given expression
   5 + 20   multiply
   25   add

2) 3(4) + 2(5)   given expression
   12 + 10   multiply
   22   add

3) 10 ÷ 2(5)   given expression
   5(5)   divide
   25   multiply

Note: In this expression the parenthesis is playing the role of a multiplication symbol and not a grouping symbol. Therefore, you can rewrite the expression as 10 ÷ 2 · 5 and continue the steps.

4) $24 \div 3 \cdot 4 + 2^5 \div 2^3 \cdot 7$ given expression
    $24 \div 3 \cdot 4 + 32 \div 8 \cdot 7$ exponents
    $8 \cdot 4 + 32 \div 8 \cdot 7$ divide
    $32 + 32 \div 8 \cdot 7$ multiply
    $32 + 4 \cdot 7$ divide
    $32 + 28$ multiply
    60 add

5) $5 + 7(12 - 9)^2$ given expression
    $5 + 7(3)^2$ simplify within parenthesis
    $5 + 7(9)$ simplify exponent
    $5 + 63$ multiply
    68 add

6) $(2)^3(5)^2(1)^2$ given expression
    $(8)(25)(1)$ simplify exponents
    200 multiply

7) $300 - 4[7 - (5 - 3)^2 + 2]^3 \div 5 \cdot 3$ given expression
    $300 - 4[7 - (2)^2 + 2]^3 \div 5 \cdot 3$ simplify within parenthesis
    $300 - 4[7 - 4 + 2]^3 \div 5 \cdot 3$ simplify exponents within brackets
    $300 - 4[3 + 2]^3 \div 5 \cdot 3$ subtract within brackets
    $300 - 4[5]^3 \div 5 \cdot 3$ add within brackets
    $300 - 4[125] \div 5 \cdot 3$ simplify exponents
    $300 - 500 \div 5 \cdot 3$ multiply
    $300 - 100 \cdot 3$ divide
    $300 - 300$ multiply
    0 subtract

8) $\dfrac{3(7 - 5)^3 \div 6 + (3 + 8)}{6 + 3^3 \div 9 \cdot 3}$ given expression

simplify the numerator and denominator following the order of operations.

simplifying the numerator

$3(2)^3 \div 6 + 11$ within parenthesis
$3(8) \div 6 + 11$ powers/exponents
$24 \div 6 + 11$ multiply
$4 + 11$ divide
15 add

simplifying the denominator

$6 + 27 \div 9 \cdot 3$ powers/exponents
$6 + 3 \cdot 3$ divide
$6 + 9$ multiply
15 add

$= \dfrac{15}{15} = 1$ divide numerator by denominator

Name _____ Date _____

EXERCISE 1.6 SET A

Find each answer. Follow the order of operations.

1) $55 - 40 + 6$

2) $12 \div 3 \cdot 2$

3) $43 + 3^4 - 2 \cdot 15$

4) $11(3) + 5(2) - 2(2)$

5) $84 \div 12 \cdot 10 \div 2$

6) $30 + 40 \div 8(2)$

7) $4(3)(1)(2)^2$

8) $5 \cdot (12 - 7 + 6 \cdot 3)$

9) $74 - (1 + 3)^3$

10) $65 - [50 - 3(8 + 2)]$

11) $11 - 5 + (13 - 2^2)$

12) $(18 + 30) \div 2 \cdot 4$

13) $122 - [3 + (4 + 3^2) \cdot 2]$

14) $3^4 \div 9 - (45 - 42)^2$

15) $2(3 + 6) - 3(15 - 6 - 9)$

16) $20 + 3[5^3 - 50 \div 2 \cdot (3 + 2)]$

17) $25 + 6\{3[5 - (3 - 2)^5 + 4(6 - 4)^3 \div 4]\}$

18) $3 + (14 - 12) \div 2 \cdot 4 - (8 - 7)^3$

19) $2^4 + \{55 - [3(12) + (2 + 4)^2 \div 6] + 1\}$

20) $[8^2 + 5 \cdot 4 \div (10 - 5)] + (12 - 5)^2$

21) $\dfrac{10 \cdot 3^3 - (5 + 15 - 10)^2}{5(4 - 2)}$

22) $\dfrac{12 + 5(6 + 3)^2 \div 3}{6 + 4^2 + (5 - 2)^3}$

EXERCISE 1.6 SET B

Find each answer. Follow the order of operations.

1) $105 + 40 - 95$

2) $48 \div 3 \cdot 4$

3) $100 - 3^4 + 2 \cdot 15$

4) $12(5) + 5(5) - 12(2)$

5) $72 \div 8 \cdot 14 \div 2$

6) $65 + 80 \div 8(5)$

7) $4^2(3^2)(1^6)(2^0)$

8) $2^3 \cdot (24 - 20 + 6 \cdot 3)$

9) $[74 - (1 + 3)^3]^2$

10) $21 + [60 - 3(8 - 2)]$

11) $11 - 5 + (13 - 2^2)$

12) $[(30 - 18) \div 2]^2 \cdot 4$

13) $92 - [39 \div (4 + 3^2) \cdot 2]$

14) $5^4 \div 25 - (35 - 33)^3$

15) $6(2 + 4) - 3[24 - 2^4 - (9 - 1^4)]$

16) $45 + 4[3^3 + 50 \div 10 \cdot (1 + 2)]$

17) $16 + 7\{12 \cdot [8 + (14 - 13)^5 - 4(7 - 5)^3 \div 4]\}$

18) $23 - (10 - 8) \div 2 \cdot 8 - (8 - 7)^3$

19) $12^2 - \{15 + [4(12) + (3 + 4)^2 \div 7] - 1\}$

20) $[9^2 + 8 \cdot 5 \div (10 - 6)] + (13 - 5)^2$

21) $\dfrac{10^2 - (5 + 15 - 10)^2}{5(4 + 2)}$

22) $\dfrac{24 + 4(5 + 3)^2 \div 16}{15 + 5^2 - (6^2 - 4^2)}$

1.7 APPLICATIONS WITH WHOLE NUMBERS

In solving math problems that are expressed in words, there are some strategies which can be used to help you. Most math instructors will use the Polya Model (developed by George Polya) for problem solving. It has been shown as an effective tool to help focus students on steps to solving word problems. Remember that the solution will not jump out at you by just reading the problem.

> **POLYA MODEL: 4 STAGES IN PROBLEM SOLVING**
> 1. Read the problem, more than once if needed. As you read the problem, ask yourself what is the given information and is that all the facts.
> 2. Devise a plan or strategy to determine the operations needed to solve the problem.
> 3. Solve the problem. Do the computation from your plan/strategy.
> 4. Check your solution. Did you answer the question? Does the answer make sense in relation to the given information?

The following are some useful problem solving strategies.
- Identify wanted, given, and needed information.
- Look for a pattern.
- Construct a table, drawing, model, or graph.
- Restate the problem in your own words.
- Solve a simpler but related problem.
- Use objects to act out the problem.
- Check for hidden assumptions.
- Account for all possibilities.
- Work backwards.

PRACTICE PROBLEMS

To lose some weight, Glen will cut out 525 Calories per day. How many Calories will Glen cut out for a week?

We will use the Poyla Model:

Step 1) given information: 525 Calories for one day
needed information: ? Calories for one week.

Since the given information is in unit terms of one day, then we need to convert the needed information of one week into the unit of days. We will use the fact that 7 days = 1 week.

Step 2) We are looking for a total using repetitive addition: 525 Cal for first day + 525 Cal for second day + ... We recognize that repetitive addition is multiplication. Therefore, the plan is to use the following computation, 525 Cal × 7 days.

Step 3) Do the multiplication, 525 × 7 = 3675 Calories

Step 4) The answer of 3675 Calories makes sense. The total should be more than the amount for one day's Calories.

Name _Sharon J Brown_   Date _8/29/08_

EXERCISE 1.7 SET A

DIRECTIONS:

Write answers with money in appropriate form.
Reduce fractions/ratios to lowest terms.
Round decimals to nearest hundredth/cent.

1) A poll of doctors in various specialties provided the following information.

Career	Number of Doctors	Average Annual Income
Family Practice Doctor	45	$142,516
Obstetricians-Gynecologists	30	$238,224
Pediatricians	38	$143,754
Cardiologists	29	$283,296
Neurologist	21	$186,653
Plastic Surgeon	42	$266,046

   a) Which doctors have average annual incomes above $180,000?
   b) How many doctors have an annual average income between $100,000 and $200,000?
   c) Find the total number of doctors whose average annual income is the two highest salaries?
   d) How much more does a plastic surgeon make than a family practice doctor?
   e) How much does a cardiologist earn, on average, each month?

2) As manager of the Clothes On store, Carrie is responsible for the 5 cashiers, 3 sales people in the men's department, 6 sales people in the women's department, 2 sales people in the shoe's department, and 1 security guard that work during her shift. How many people is Carrie responsible for in all?

3) A nurse in the operating room earns $865 per week for 40 hours worked, plus $230 for 4 hours of overtime. What is the nurse's total income for the week?

4) As a publishing company's sales representative you are expected to travel to meet with clients. You had the following expenses during one week. Calculate your total expenses for the week.
   lunches $40   gas $50   tolls $5

5) As a fashion consultant you earned $6344 last month. You also earned $500 last month by selling one of your own designs to a boutique. You spent $345 on fabric, $88 on buttons and $25 on thread.
   a) What were your total expenses?
   b) What was your total income for last month?
   c) Did you make a profit on selling one of your own designs?

6) Last month your take-home pay as a hairdresser was $1988. You also worked part-time as a bartender and took home $857. Your expenses for the month were as follows:

rent	$650	car loan	$299	cell phone	$59
utilities	$150	car insurance	$110	credit cards	$125
groceries	$150	other	$320		

   a) What was your total take-home pay last month?

   b) What were your total expenses for last month?

   c) Did you have enough money left to purchase two plane tickets for $399 each for a four day weekend get away?

7) Yolanda Parta's monthly take-home pay as a nurse is $3750. Yolanda's husband's monthly take-home pay as a science teacher is $4583. He also works part-time as a mechanic and earns $600 per month. Their expenses for each month are the following:

mortgage	$1650	car loan	$359	utilities	$450
car insurance	$105	cell phone	$69	credit cards	$265
groceries	$350	day care	$500	entertainment	$500
other	$200				

   a) What is the combined monthly take-home pay of the Parta family?

   b) What are the Parta's monthly expenses?

   c) Their son, Richard, must get braces which costs $3455. The medical insurance only covers $1800. Is their enough money left after paying the monthly bills to cover the Parta's portion of the dentist bill?

8) You have budgeted $650 this month to purchase furniture. You spent $290 on a floor lamp. How much remains in your furniture budget?

9) For the month of August you budgeted $75 for groceries and $200 for eating out. Your best friend called to let you know that he was coming for a visit. You increased your grocery budget to $150 and eating out budget to $350. How much more money are you budgeting in all now that your best friend is coming for a visit?

10) Wanda received a promotion as supervisor of data entry. With this new position Wanda feels it is important to be well-dressed. Wanda would like to buy a new suit on sale for $299. This month Wanda already spent $55 of the $150 budgeted for miscellaneous expenses each month.

   a) Before buying the suit, how much money remains in the miscellaneous budget?

   b) How much additional money will Wanda need to buy the suit?

   c) Will the next two month's miscellaneous budget be enough to pay for the remaining cost of the suit?

11) On March 1 Ernie's checking account balance was $1235. On March 2 Ernie's paycheck was deposited and the balance was $2349. How much was Ernie's paycheck?

12) On May 1st your checking account balance was $77. On May 15th you deposited your paycheck of $1344. On May 16th you wrote checks in the amounts of $455, $69, and $580. On May 20th you deposited $200. On May 31st you deposited your paycheck of $1344. On June 1st you wrote checks in the amounts of $750, $150, and $350.

   a) What is the total amount of checks written in May and June?

   b) What is the total amount of deposits made in May?

   c) What was your account balance on May 25th?

   d) What was your account balance on June 2nd?

13) To lose some weight, Glenda will cut out 450 Calories per day. How many Calories will Glenda cut out for a week?

14) Doctors have recommended that children between the ages of 9 and 12 need approximately 2500 Calories per day in their daily diets. How many Calories will that be in a year?

15) You work off 300 Calories per hour by power walking, 450 Calories per hour by swimming, 250 Calories per hour on the stationary bike, and 320 Calories taking a body toning class. During one week you power walked for 3 hours, rode the bike for 3 hours, and took the body toning class twice. How many Calories did you burn that week from doing the various exercises?

16) You want to lose 2 pounds per week. After some research you conclude you will need to consume 4500 Calories less per week to lose the pounds. Your plan each day is to consume 100 Calories less at breakfast, 300 Calories less at lunch and 250 Calories less at dinner. Will your plan be good enough for you to lose the 2 pounds?

17) A box of individual sized bags of potato chips has a dozen bags. Mrs. Hague buys two boxes of potato chips to distribute to all her classes. How many bags of potato chips did Mrs. Hague buy?

18) Sleep Motel advertises $39 per night for a room with cable television. Tony thinks this is a good deal and books a room for 15 nights. How much is Tony's total bill?

19) You will be reimbursed by your company for meals and transportation while you attend a conference. You spent about $40 per day for meals and $8 in taxi fares each day. The conference lasted 5 days. What is the total amount your company will reimburse you?

20) The annual trip for the Girl Scouts this year is to the Bronx Zoo in New York and will be paid from their annual trip fund. Admission for the Bronx Zoo is $8 per adult (13 years and above) and $6 per child (under 12 years old). All day parking is $15 per bus. There will be 45 girls under 12 years old, 15 girls older than 12 years old, and 12 adult chaperones.
   a) What is the total amount for admission to the Bronx Zoo?
   b) If each bus holds 35 people, how many buses will be needed for this trip?
   c) What is the total amount for admission and parking?

21) A travel agent offers the Let's See the World Club a 2-week package to tour Russia for $2100 per person or $150,000 for a group of 100 or fewer persons. Which will be cheaper for a group of 75 club members, the individual rate or group rate?

22) The cost of a cab ride is $5 plus 15¢ per mile. On Monday, Rosetta took a cab from Hoboken to Bloomfield whose distance is 15 miles and gave the driver a $5 tip. What was the total cost for Rosetta's cab ride?

23) An 8-pound bag of dog food costs $32. What is the unit price (price per pound)?

24) A 12-can package of orange soda costs $9. What is the cost per can?

25) Store #1 sells a box of 48 assorted Christmas cards costing $8.16 (816¢). Store #2 sells a box of 58 assorted Christmas cards costing $9.28 (928¢).
   a) What is the price per card for store #1?
   b) What is the price per card for store #2?
   c) Which store offers the better bargain?

26) Your health insurance premium has been increased to $216 per year. You will pay this amount in monthly installments. How much will each monthly payment be?

27) Your annual salary is $45032. How much is your weekly salary?

28) In Laws County, the property tax on a house worth $345,000 is $7200 per year. A home-owner will pay this amount semi-annually. How much is each payment?

29) Your company pays three-quarters of the dental plan and the remainder is paid equally by the 24 employees. The total annual cost for the employee's portion of the dental plan is $7320. How much will each employee pay per year?

30) For each $1000 of life insurance purchased, the annual premium is $49. If Leslie purchases a $35,000 life insurance policy, how much is her yearly payment?

31) For each $1000 of life insurance purchased, the annual premium is $99. Amil decides to purchase a $100,000 life insurance policy.
   a) How much is Amil's annual premium?
   b) If Amil decides to pay quarterly, how much is his quarterly payment?

32) For each $1000 of life insurance purchased, the annual premium is $120. You decide to purchase a $500,000 life insurance policy.
   a) How much is your monthly payment?
   b) If you decide to pay quarterly, how much is your quarterly payment?

33) The Yardley's car insurance policy has been decreased because they now park the car in a garage. The cost of the new policy is $1200 per year.
   a) What is the Yardley's new monthly payment?
   b) The new cost has decreased the Yardley's monthly payment by $50. What was the Yardley's old cost per year?
34) Sharif bought his textbooks at the campus bookstore. He spent $75 for his math text, $110 for his history text, $18 each for 3 novels, and $115 for his accounting text. What was the total cost if sales tax was $21?
35) Bake For You Bakery sells Italian candy for $8 per pound. One pound of the candy contains 85 pieces.
   a) A wedding favor will include 5 pieces of the Italian candy. If Elisa needs 187 wedding favors, how many pounds of the Italian candy will she need to buy?
   b) How much will Elisa pay for the Italian Candy?
36) For last summer's vacation the Granger family did a car trip along the East coast. When they started their trip, the odometer read 67,789 miles. At the end of their trip the odometer read 69,014 miles.
   a) How many miles was the Granger's car trip?
   b) The car would get approximately 25 miles per gallon of gas. How many gallons of gas did the car use for the trip?
37) In a recent sweepstakes there were ten winners who won varying amounts. One person won $500,000. Two people won $100,000 each. Two people won $10,000 each. Five people won $1000 each. What is the total amount of money paid out by the sweepstakes?
38) Belinda drinks 2 cups of herbal tea every day. She uses 3 teaspoons of reduced-calorie sugar for every cup. How many teaspoons of sugar does Belinda use in a year?
39) Greg needs 12 pieces of 2" by 4" oak wood. Each piece of wood is 6 inches in length. How many total inches of wood does Greg need?
40) The plumber ordered 4 pieces of three-inch pipe at $6 per inch and 8 pieces of five-inch pipe at $11 per inch.
   a) How many total inches of three-inch pipe were ordered?
   b) How many total inches of five-inch pipe were ordered?
   c) What was the total cost of the pipes?
41) If you drive 65 miles per hour, how many miles will you travel in 3 hours?
42) In a 5-day road trip, Martina drove 2550 miles over a total of 50 hours. What was her average speed (miles per hour)?
43) When Julia purchased the Lady Spa in Los Angeles, she agreed to pay 206 payments of $3,355 each. How much did she agree to pay in total?
44) Malik received a federal grant of $49,350 to distribute for computer support to the 21 sixth grade classes in his school district. How much money did each class get?
45) Omar had $6218 in federal income taxes deducted from his paychecks last year. When Mark computed his federal income tax, his tax was $5929. How much will he receive in a refund from the government?
46) Mai is an efficiency expert hired by NYC to study the Finance Department. She determined that using the tax computers, the City can print 11 bills per minute. How long will it take the City to print the 3,883 bills for the borough of The Bronx?
47) For the past ten years, the town of Gringoll has charged its homeowners $25 every month for garbage pick up. How much total was each homeowner charged for garbage pick up?
48) Every Tuesday Josephina's Pizzeria offers the deal "buy 2 large pizza pies, get a third pie free". If you buy two large pizza pies at $9 per pie, what is the actual cost per pizza pie?
49) "Speed Up" internet service offers $12 per month for service. "This Is Quick" internet service says they are the better deal for $150 per year. Is this true?

EXERCISE 1.7 SET B

DIRECTIONS

Write answers with money in appropriate form.
Reduce fractions/ratios to lowest terms.
Round decimals to nearest hundredth/cent.

1) In August, a new company, Super Tutor, hired 17 math tutors, 15 English tutors, 8 chemistry tutors, 5 biology tutors, 10 computer tutors, and 5 psychology tutors. In December, Super Tutor re-evaluated their needs for tutors and made the following changes; hired 5 more math tutors, 5 more English tutors, and 2 more computer tutors; let go 4 chemistry tutors and 2 psychology tutors.

   a) How many total tutors were hired in August?
   b) How many total tutors were let go?
   c) How many more math tutors were employed than English tutors in December?
   d) How many chemistry tutors are left?

2) A high school student earns $200 per month as a private math tutor. He also earns $100 per month as a cashier in his Dad's deli. What is his total monthly income?

3) The Let's Practice the SAT company hired several new people this year in several different areas as shown in the table below.

# of employees	Job Description
23	Create math problems
12	Create English questions
13	Proofread math problems
6	Proofread English questions
25	Create answer key for math problems
25	Create answer key for English questions

   a) How many employees were hired this year?
   b) How many employees were hired in the math area?
   c) How many more employees were hired in creating an answer key for English questions than creating English questions?

4) Your take-home pay for the month included:

| human resource consultant | $4345 | | part-time wedding photographer | $1400 |

Your average expenses for the month included:

rent	$1200	dry-cleaning	$100	entertainment	$300
utilities	$180	car loan	$310	groceries	$250
cell phone	$49	college loan	$188	credit card	$75

    a) What was your total take-home pay for the month?

    b) What were your total expenses for the month?

5) During the winter, Herman operates a part-time business in snow removal for commercial businesses. In 2004, Herman's four employees each earned $4500. Also, Herman spent $1066 in total in equipment repairs, $400 on rock salt, and $250 on advertising his business. Herman had a total of 25 clients in which each client paid $230 for 5 snow removals each in 2004.

    a) What was Herman's income?

    b) What were Herman's expenses?

    c) How much profit did Herman make with his part-time business?

6) Lenny is a graduate assistant at a local college and earns $880 a month. Lenny also is a waiter on the weekends and earns about $350 a month. His expenses for the month are $550 for rent, $100 on groceries, $50 on transportation, $150 for utilities, $49 for a cell phone, and $200 on entertainment.

    a) What is Lenny's take-home pay for the month?

    b) What are Lenny's expenses for the month?

    c) Does Lenny have enough money left to purchase two concert tickets at $69 each as a gift for his girlfriend's birthday?

7) On the weekends you run a hot dog cart outside of Grand Central Station in NYC and must replenish your supplies each weekend. On the first weekend of April you earned $235 and spent $120. On the second weekend of April you earned $330 and spent $190. On the third weekend of April you earned $190 and spent $100. On the fourth weekend of April you earned $440 and spent $210.

    a) What is the total amount you earned in April from your hot dog cart?

    b) What is the total amount you spent in April on supplies for your hot dog cart?

8) The local chapter of the Knitting Club budgeted $500 for yarn. Of this amount, $260 has already been spent. How much remains in the budget?

9) During the winter break session of Pam's first year of college, Pam earned $330 babysitting and $140 cleaning a house. The day before the semester began Pam bought three of her textbooks for the amounts of $95, $105, and $65. How much is left from her winter break earnings to buy the remainder of her textbooks?

10) At the end of last year, your savings account balance was $4800. In the new year, you needed to withdraw money to pay for a new roof on your home. Your new balance was $1044. How much did the new roof cost?

11) On Friday you wrote four checks in the amount of $289, $100, $45, and $378. Your checking account balance was $915. How much money is left in the checking account?

12) Sandra deposited a check for $250 in her savings account. Her savings account balance before the deposit was $1095. She then needed to withdraw $500 for car repairs. What is Sandra's balance in her savings account?

13) On Wednesday you deposited $355 in your checking account. Before the deposit your balance was $119. On Thursday you wrote checks for $102 and $175. On Friday the bank charged your account $7 for service fees. On Saturday morning you want to write a check for $75. Do you have enough in your checking account to write the check?

14) A nutrition label on a box of graham crackers states that one graham cracker is 16 Calories. How many Calories will 8 graham crackers be?

15) Jack's plan is to lose 2 pounds per week. If Jack succeeds in losing weight at that rate, then how much total weight will he lose in one year?

16) The doctor recommends Chantal consume a total of 10,500 Calories per week. For the first week, Chantal consumed 1450 each day. Did Chantal consume more or less for the week than what the doctor recommended?

17) For breakfast, Nelly had two slices of wheat toast with two teaspoons of peach jam, an 8-ounce glass of orange juice, a cup of coffee with two teaspoons of sugar, and a pear. According to the book, *Calorie Counting*, a slice of wheat bread has 95 Calories, a 4-ounce glass of orange juice has 75 Calories, a teaspoon of jam has 45 Calories, a teaspoon of sugar has 65 Calories, and a pear has 50 Calories. What is the total amount of Calories consumed by Nelly for breakfast?

18) You would like to lose 2 pounds each week. You have decided to exercise (burn 350 Calories per day) and stop drinking 8-ounce cups of cappuccinos (8 ounces contains 300 Calories). To lose one pound you will need to have a loss of 2000 Calories.
   a) How many Calories will you burn if you exercised for 5 days during the week?
   b) How many Calories do you cut out if you do not drink 8 cups of cappuccinos during the week?
   c) Will you lose two pounds per week?

19) A 5-pound bag of onions has approximately 8 onions. Chef Pierre buys 12 bags of onions from the local farmer's market. How many onions did Chef Pierre buy?

20) The high school chorus is planning their annual trip to the ski resort. This year the resort is charging $35 per day per guest for the room and $12 per guest per day for ski rental. There are 18 students and 4 chaperones going on the trip and all of them need to rent skis except for two people. What is the total cost of the 4-day annual trip?

21) Last year Walter planned the Crearie family reunion on a cruise to the Bahamas. The cruise tickets were $1500 per adult, $1250 per young adult (ages 13–17), and $990 per child (ages 1–12). Flight tickets to Tampa, Florida where the cruise began were $225 per person. Thirty-five adults, ten young adults, and 8 children came to the Crearie family reunion. What was the total bill for the cruise and flight?

22) The Ecology Club has $2355 in their budget for guest speakers. The Club would like to invite the town's Administrator of Environmental Policies to speak at a rally on Earth Day. The expenses associated with the invitation are $1400 for the speaker's fee, $18 per person for lunch (55 people are expected to attend), and $200 for advertising the event. Does the Ecology Club have enough money in their budget?

23) Your company will reimburse you for driving to see clients. The amount is 25¢ per mile. In February, you traveled 40 miles. How much did the company pay you?

24) Carol and her four children will drive to visit her twin brother Carl during the summer. Carl lives 2100 miles away. To rent a van the cost will be $200 plus 5¢ per mile. What is the round trip cost for renting the van?

25) A dozen eggs costs $3.60. What is the cost per egg?

26) Parking your car at work costs $480 per year? What is the cost of parking per month?

27) You need 150 individualized ice cream cups for the annual company picnic. Brand A costs $3.60 for six cups in a box. Brand B costs $13.20 for twenty-four cups in a box.
   a) What is the price per cup of Brand A?
   b) What is the price per cup of Brand B?
   c) How much will 150 ice cream cups of the cheaper brand cost?

28) You have turned 25 and your car insurance has been decreased to $1180 per year. You will pay this amount in quarterly installments. How much will each quarterly payment be?

29) You received a notice from your condominium maintenance committee. Your maintenance fees for the next 3-year period will be $1260. You decide to pay monthly. How much will your monthly payments be?

30) The annual property tax on your home was $6500 for Alps County. The state re-assigned county lines. Your home now is in Hills County whose annual property tax is $7892. How much more will you pay each month for property taxes on your home?

31) For each $1000 of life insurance purchased, the annual premium is $119. If Howard purchases a $25,000 life insurance policy, how much is his yearly payment?

32) For each $1000 of life insurance purchased, the annual premium is $105. Casey decides to purchase a $250,000 life insurance policy.
   a) How much is Casey's annual premium?
   b) If Casey decides to pay quarterly, how much is her quarterly payment?

33) Mr. & Mrs. Wheeler have narrowed their choices to two houses. The house in Plattsville is assessed at $350,000 and the property tax is $12 per thousand per year. The house in Olaf is assessed at $400,000 and the property tax is $9 per thousand per year.
   a) How much property tax will Mr. & Mrs. Wheeler pay per year if they live in Plattsville?
   b) How much property tax will Mr. & Mrs. Wheeler pay per year if they live in Olaf?
   c) How much will Mr. & Mrs. Wheeler save each month if they live in the town with the cheaper property tax?

34) You decided to purchase the extended warranty on your car. The warranty costs $1650 for a 5-year period.
   a) How much did the warranty cost per year?
   b) How much did the warranty cost per month?
   b) How much did the warranty cost per week?
   c) How much did the warranty cost per day?

35) A crate can hold 100 boxes. Each box can hold three dozen bottles of orange juice. At a local supermarket the delivery people had to fill the crate 4 times to transfer the orange juice from the truck to the refrigerator. How many bottles of orange juice were delivered?

36) The cost for gas during one month was $2 per gallon. During that month you filled up your gas tank 5 times with the following amounts; 10 gallons, 8 gallons, 12 gallons, 9 gallons, and 12 gallons. How much did you spend on gas that month?

37) At the end of her shift, Kelly had two $100 bills, thirteen $20 bills, twenty $10 bills, eight $5 bills, and thirty-three $1 bills in her register. How much money did Kelly have at the end of her shift?

38) A building contains five office suites with 484 square feet of floor space each, a conference room with 225 square feet of floor space, two bathrooms with 108 square feet of floor space each, and a kitchen with 180 square feet of floor space. What is the total floor space of the building?

39) Valdimir bought a 36-inch television for $699. He made a down payment for $275. The remainder of the cost will be made in monthly payments of $53. In how many months will Valdimir be done paying for his new television?

40) The Fabric Are We store advertised silk fabric at $35 per yard, cotton fabric at $12 per yard, and rayon fabric at $6 per yard. Federico needed 10 pieces of silk fabric at 10 yards each, 5 pieces of cotton fabric at 10 yards each, and 12 pieces of rayon fabric at 8 yards each. How much did Federico spend for the fabric he needed?

41) If Justin and Brittany each drove 4 hours at 70 miles per hour, how many miles did they drive?

42) Dwayne's new speedboat goes 125 miles per hour. In one day, Dwayne drove for 5 hours. How many miles did Dwayne drive?

43) A mother whale weighs 82 times as much as her baby. The baby whale weighs 38 pounds. How much does the mother weigh?

44) Kim earned $30,078 in 18 months working as a teacher's assistant while she studied for her degree. How much did she earn, on average, each month?

45) Nicole's unit of the Red Cross collected 32,500 units of blood to be distributed to local chapters at a rate of 250 units each. How many local chapters could receive blood?

46) Tickets to Neptune College's annual Reach for The Sky Scholarship Dinner were $12 per student, $18 per faculty/staff person, and $24 per alumni. What was the total revenue for the Dinner if 58 students, 77 faculty or staff, and 45 alumni attended?

47) Julien drove 2256 miles in his Chevy Blazer to visit his girlfriend Amie in Texas. Julien's truck averages 24 miles per gallon of gasoline. Gas is $2 per gallon. What was the total amount he spent for gasoline?

48) The top of a mountain is 1470 feet below a cloud. The mountain is 12,805 feet above sea level. An eagle flies 7516 feet above sea level. How far below the cloud is the eagle?

49) Jarred is five years older than Jenna. Jenna is twice as old as Jude. Jude is twenty-three years younger than Jackie. Jackie is forty years old. How old is Jarred?

50) One hundred fifty million people pledged, on average, $50 for hunger relief. How much money was donated for hunger relief?

51) "You Need Furniture" advertises no payments for a year. However, the fine print states if full payment is not made within the year, then for each day the full payment is not made, the customer will be charged $10. Henrietta bought a bedroom set for $4500 on January 3, 2004. On January 3, 2005, Henrietta had paid $2000.

  a) How much remains for Henrietta to pay?

  b) Henrietta sets up a plan for herself to pay the remaining amount with payments of $500 on January 31, February 15, February 28, March 15, and March 31. How much will Henrietta be charged for not paying within the year?

CHAPTER 2

2.1 ADDITION AND SUBTRACTION WITH SIGNED NUMBERS

Through our study of arithmetic, we worked with numbers that were larger than zero. These numbers are called positive numbers. They are sometimes written with a "+" sign, such as +5 instead of just 5. Here the "+" sign is not used to signify the operation of addition. It is used to indicate that the number is larger than zero—a positive number.

Before we can proceed into an introduction of algebra, we must learn about another group of numbers called negative numbers. You have probably encountered these numbers before.

If the temperature is 12° below zero, we say it is minus twelve degrees ($-12°$). If you write a check for $50.75, when you have only $20.00 in your checking account, you would be overdrawn and have a balance of $-$30.75.

These numbers that have values less than zero are the negative numbers. They are the opposite of positive numbers. The "$-$" sign used with these numbers does not signify the operation of subtraction. It means you have a number that is less than zero. These two types of numbers, positive and negative numbers, are called **signed numbers**, since they are written with either a positive sign ($+$) or a negative sign ($-$).

The number zero separates the positive numbers and the negative numbers. It is neither positive or negative. It is the only neutral number in our number system. We can show the relative values of signed numbers by placing them along a line called a **number line.**

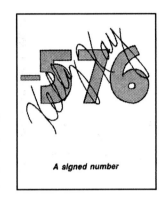

A signed number

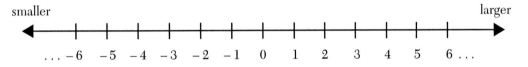

You can use the number line to compare signed numbers. *As you go to the right along the number line, the numbers get larger; and as you go to the left along the number line, the numbers get smaller.*

For example: $3 > -2$ 3 is greater than -2, since 3 is farther to the right on the number line than -2.

$-4 < 0$ -4 is less than 0, since -4 is farther to the left on the number line than 0.

$-6 > -10$ -6 is greater than -10, since -6 is farther to the right on the number line than -10.

$-12 < -7$ -12 is less than -7, since -12 is farther to the left on the number line than -7.

Fractional and decimal numbers can also be placed on the number line by placing them between the positive and negative whole numbers.

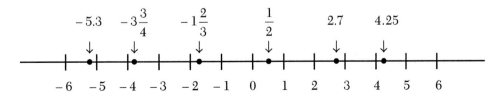

1.) Pages 009–018, 037–042, 045–047, 052–059, 062–077, 080–087, 115–127, 129–132, 141–143, 147–152, 184–187, 199–203, 206–209, from *Math for College Students: Arithmetic with Introductions to Algebra and Geometry*, 5th Edition by Ronal Straszkow. © 1997 by Kendall/Hunt Publishing Company. Used with permission.

Now that you understand what is meant by signed numbers, you can learn to perform the basic operations on these numbers. The first operation is the addition of signed numbers. To help explain how this is done, we will use a number line.

> **ADDING SIGNED NUMBERS CAN BE ILLUSTRATED ON A NUMBER LINE AS FOLLOWS:**
>
> 1. Adding a positive number is the same as moving that many units in the positive direction.
> 2. Adding a negative number is the same as moving that many units in the negative direction.

Example 1

$5 + (-3) = ?$

That problem suggests that you start at 5 on a number line and go 3 units in the negative direction.

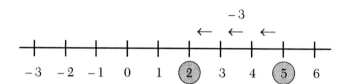

So, $5 + (-3) = 2$

Example 2

$-7 + 4 = ?$

Start at -7 on a number line and go 4 units in the positive direction.

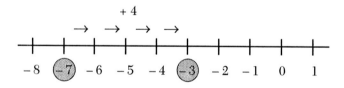

So, $-7 + 4 = -3$

Example 3

$-2 + (-6) = ?$

Start at -2 on a number line and go 6 units in the negative direction.

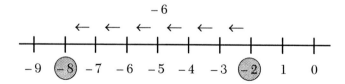

So, $-2 + (-6) = -8$

If the numbers to be added had many digits, such as $-576 + 329$, using a number line to find the answer would really be impractical. Because of that difficulty, let us consider a method that will enable us to add signed numbers more quickly than using a number line.

Every number has two parts: a "+" or "−" sign, and a number part.

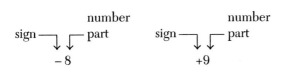

If a number has no sign, it is understood to be a positive number.

$$16 = +16$$

> **RULES FOR ADDING SIGNED NUMBERS**
> 1. If the numbers have the same signs, add their number parts and use that same sign as the sign of the answer.
> 2. If the numbers have different signs, subtract their number parts and use the sign of the larger number part.

Example 4

$-5 + (-8) = ?$

The numbers have the same signs.
Add the number parts.

$$\begin{array}{r} 5 \\ \underline{8} \\ 13 \end{array}$$

Since both are negative, the answer is negative.
So, $-5 + (-8) = -13$.

Example 5

$-5 + 8 = ?$

The numbers have different signs. Subtract the number parts.
Subtract the smaller number part (5) from the larger (8).

$$\begin{array}{r} 8 \\ \underline{5} \\ 3 \end{array}$$

Since the larger number part (8) is positive, the answer is positive.

So, $-5 + 8 = 3$.

You could also obtain the answer by picturing movement on a number line. If you start at -5 and move 8 units in the positive direction you will end up at $+3$.

Example 6

$-75 + (-46) = ?$

The numbers have the same signs.
Add the number parts.

$$\begin{array}{r} 75 \\ \underline{46} \\ 121 \end{array}$$

Since both are negative, the answer is negative.

So, $-75 + (-46) = -121$

Example 7

27 + (−52) = ?

The numbers have different signs. Subtract the number parts.
Subtract the smaller number part (27) from the larger (52).

$$\begin{array}{r} 52 \\ \underline{27} \\ 25 \end{array}$$

Since the larger number part (52) is negative, the answer is negative.

So, 27 + (−52) = −25.

Example 8

−35 + 19 = ?

The numbers have different signs. Subtract the number parts.
Subtract the smaller number part (19) from the larger (35).

$$\begin{array}{r} 35 \\ \underline{19} \\ 16 \end{array}$$

Since the larger number part (35) is negative, the answer is negative.

So, −35 + 19 = −16.

In the previous section, we learned to add signed numbers by using either a number line or the rules for adding signed numbers. In order to perform subtraction on signed numbers, we will learn to change a subtraction problem into an addition problem. Consider the problems 5 − 2 = ? and 5 + (−2) = ?.

---

5 − 2 = 3 and 5 + (−2) = 3

Those examples are different ways to represent the same problem. The first one, 5 − 2, is a subtraction problem, and the second one, 5 + (−2), is an addition problem. This points out how to subtract signed numbers.

RULES FOR SUBTRACTING SIGNED NUMBERS

1. Change the sign of the number being subtracted.
2. Change the subtraction to addition.

That is, for any numbers represented by *A* and *B*:

$$A - B = A + (-B)$$

For example:
$13 - 7 = 13 + (-7) = 6$
$6 - 9 = 6 + (-9) = -3$
$-4 - 6 = -4 + (-6) = -10$
$75 - 97 = 75 + (-97) = -22$
$-37 - 21 = -37 + (-21) = -58$

Notice that each subtraction was changed into addition of signed numbers, then the rules for addition were applied.

PROBLEMS

1) $15 - 23 = ?$　　　Answer: -8

$$15 - 23 = 15 + (-23) = -8$$

2) $-30 - 12 = ?$　　　Answer: -42

$$-30 - 12 = -30 + (-12)$$
$$= -42$$

You will notice that the "$-$" sign is used in different ways.

1. To represent subtraction. $12 - 8$ means 12 subtract 8.

2. To signify a negative number. -7 indicates a number that is less than zero—a negative number.

There is yet another way to use the "$-$" sign. It is sometimes used to indicate the inverse or opposite of a number.

$-(+8)$ indicates the opposite of a $+8$, which is -8.

$$-(+8) = -8$$

$-(-3)$ indicates the opposite of a -3, which is $+3$.

$$-(-3) = +3 = 3$$

This usage of the "$-$" sign gives us a way to determine the result of a number that has two negative signs in front of it.

$$-(-7) = +7 = 7$$
$$-(-35) = +35 = 35$$
$$-(-103) = +103 = 103$$

That property of signed numbers should be familiar to us, since, even in everyday English, two negatives give a positive statement.

For example: "I'm *not mis*behaving!" implies the positive meaning that you are behaving.

Similarly, $5 - (-2)$ becomes $5 + 2$, which equals 7.

$$8 - (-7) = 8 + 7$$
$$= 15$$

$$-17 - (-8) = -17 + 8$$
$$= -9$$

$$-67 - (-67) = -67 + 67$$
$$= 0$$

You will notice that this is consistent with the rules for subtracting signed numbers as stated on the previous page.

PROBLEMS

3) $28 - (-12) = ?$　　　Answer: 40

$$28 - (-12) = 28 + 12 = 40$$

4) $-8 - (-15) = ?$　　　Answer: 7

$$-8 - (-15) = -8 + 15 = 7$$

5) $-20 - 11 = ?$　　　Answer: -31

$$-20 - 11 = -20 + (-11)$$
$$= -31$$

6) $-43 - (-31) = ?$　　　Answer: -12

$$-43 - (-31) = -43 + 31$$
$$= -12$$

Name _____ Date _____

EXERCISE 2.1 SET A

Find each answer.

1) $25 + 15$

2) $25 + (-15)$

3) $-25 + (-15)$

4) $-25 + 15$

5) $15 + (-25)$

6) $(-15) + 25$

7) $12 + (-20)$

8) $-45 + 10$

9) $19 + (-33)$

10) $-66 + (-11)$

11) $124 - 154$

12) $16 - 32$

13) $-52 - 22$

14) $-101 - 81$

15) $-30 + 17$

16) $28 + (-27)$

17) $-3 + (-3)$

18) $60 + (-61)$

19) $-18 + (-10) + (-4)$

20) $21 - 12 + (-7)$

21) $36 + (-42) + 6$

22) $-72 - 12 - 20$

23) $-85 - 35 + (-2) + 90$

24) $77 + (-84) + 7$

25) $16 - (-12)$

26) $12 - (-16)$

27) $-16 - (-12)$

28) $-12 - (-16)$

29) $48 - (-8) + 22$

30) $-19 - 14 - (-9)$

31) $27 + (-47) - (-20)$

32) $-(-5) - (-6) + 12$

33) $100 + (-50) - (-25) - 15 - 10$

34) $-98 - (-14) + (-6) - 8 + 28$

35) $-2 - (-5) + (-8) - 7 - (-4) + 3$

36) $9 + (-6) - (-1) - 2 - 7 - (-8)$

EXERCISE 2.1 SET B

Find each answer.

1) $3 + 11$

2) $11 + (-19)$

3) $-15 + (-13)$

4) $-16 + 15$

5) $25 + (-18)$

6) $(-17) + 25$

7) $22 + (-26)$

8) $-54 + 12$

9) $16 + (-34)$

10) $-44 + (-44)$

11) $213 - 312$

12) $26 - 42$

13) $-93 - 13$

14) $-110 - 56$

15) $-50 + 11$

16) $18 + (-27)$

17) $-13 + (-13)$

18) $21 + (-24)$

19) $-19 + (-9) + (-3)$

20) $7 - 14 + (-7)$

21) $41 + (-36) + 22$

22) $-43 - 13 - 33$

23) $-58 - 81 + (-9) + 20$

24) $99 + (-77) + 9$

25) $26 - (-16)$

26) $15 - (-18)$

27) $-18 - (-12)$

28) $-32 - (-10)$

29) $8 - (-8) + 2$

30) $-24 - 12 - (-4)$

31) $36 + (-46) - (-50)$

32) $-(-8) - (-9) + 17$

33) $124 + (-24) - (-54) - 44 - 100$

34) $-18 - (-104) + (-66) - 80 + 28$

35) $-3 - (-4) + (-5) - 6 - (-7) + 8$

36) $2 + (-3) - (-8) - 9 - 4 - (-5)$

2.2 MULTIPLYING AND DIVIDING SIGNED NUMBERS

MULTIPLYING SIGNED NUMBERS

To arrive at a method for multiplying signed numbers, we must consider four possibilities.

1. **A positive number times a positive number.**

$$(+3) \bullet (+4) = ?$$

The answer will be a positive number, 12.

2. **A positive number times a negative number.**

$$(3) \bullet (-4) = ?$$

In our first discussion of multiplication in Section 1.5, we saw that multiplication is actually a way to represent repeated addition. $3 \bullet (-4)$ means that you have three -4's; that is, $(-4) + (-4) + (-4)$. The answer to that is -12.

$$\text{So, } 3 \bullet (-4) = -12$$

The result of multiplying a positive times a negative number is a negative number.

3. **A negative number times a positive number.**

$$-3 \bullet 4 = ?$$

By similar reasoning, the result of a negative number times a positive number is a negative number.

$$-3 \bullet 4 = -12$$

4. **A negative number times a negative number.**

$$-3 \bullet (-4) = ?$$

Consider the list below:

$-3 \bullet 3 \quad = -9$
$\quad\quad\quad\quad\quad\quad +3$
$-3 \bullet 2 \quad = -6$
$\quad\quad\quad\quad\quad\quad +3$
$-3 \bullet 1 \quad = -3$
$\quad\quad\quad\quad\quad\quad +3$
$-3 \bullet 0 \quad = 0$
$\quad\quad\quad\quad\quad\quad +3$
$-3 \bullet (-1) = \;?$
$\quad\quad\quad\quad\quad\quad +3$
$-3 \bullet (-2) = \;?$
$\quad\quad\quad\quad\quad\quad +3$
$-3 \bullet (-3) = \;?$
$\quad\quad\quad\quad\quad\quad +3$
$-3 \bullet (-4) = \;?$

As we proceed down the list, each new answer can be obtained by adding 3 to a previous answer.

If you continue the pattern, you will find that:

$$-3 \bullet (-4) = 12$$

A negative number times a negative number is a positive number.

The results can be summarized as follows:

$$(+) \bullet (+) = (+)$$
$$(-) \bullet (-) = (+)$$
$$(+) \bullet (-) = (-)$$
$$(-) \bullet (+) = (-)$$

Notice, if the numbers being multiplied have the same signs, the answer will be positive. If they have different signs, the answer will be negative.

DIVIDING SIGNED NUMBERS

You will be pleased to know that there are no new rules for dividing signed numbers. See page 79 on dividing fractions. Dividing by a number will give the same result as multiplying by an appropriate fraction. For example,

$$8 \div 2 = 8 \bullet \frac{1}{2} = 4 \quad \text{and} \quad 12 \div 3 = 12 \bullet \frac{1}{3} = 4$$

Since division can be changed into multiplication, the rules to obtain the sign of the answer in division are the same as those in multiplication.

RULES FOR MULTIPLYING AND DIVIDING SIGNED NUMBERS

1. Multiply or divided the number parts.
2. If the numbers have the same signs, the answer will be positive.
3. If the numbers have different signs, the answer will be negative.

PROBLEMS

Perform the indicated operations:

Answers:

1) $-7 \bullet 5 = ?$ 1) -35

2) $-9 \bullet (-6) = ?$ 2) 54

3) $40 \div (-8) = ?$ 3) -5

4) $+27 \div 3 = ?$ 4) 9

5) $4 \bullet (-7) = ?$ 5) -28

6) $-45 \div (-15) = ?$ 6) 3

7) $-12 \bullet 1 = ?$ 7) -12

8) $24 \div (-4) = ?$ 8) -6

Name _____ Date _____

EXERCISE 2.2

Perform the indicated operations.

_____ 1) $5 \cdot (+3)$ _____ 2) $6 \cdot (+4)$

_____ 3) $3(-7)$ _____ 4) $2(-9)$

_____ 5) $-4 \cdot 8$ _____ 6) $-5 \cdot 6$

_____ 7) $(-8)(-6)$ _____ 8) $(-7)(-5)$

_____ 9) $-9 \cdot (-8)$ _____ 10) $-7 \cdot (-9)$

_____ 11) $\dfrac{+36}{9}$ _____ 12) $\dfrac{+27}{3}$

_____ 13) $\dfrac{-40}{5}$ _____ 14) $\dfrac{-42}{7}$

_____ 15) $\dfrac{56}{-8}$ _____ 16) $\dfrac{63}{-9}$

_____ 17) $\dfrac{-72}{-9}$ _____ 18) $\dfrac{-32}{-8}$

_____ 19) $\dfrac{-25}{-5}$ _____ 20) $\dfrac{-36}{-6}$

_____ 21) $\dfrac{143}{-11}$ _____ 22) $\dfrac{144}{-12}$

2.3 EXPONENTS AND ORDER OF OPERATIONS WITH INTEGERS

In a previous section, simplifying expressions with exponent notation was demonstrated. Now that we have the sign rules for the operations, we need to revisit some examples.

RECALL

EXPONENT NOTATION

$a^b = a \cdot a \cdot a \cdot a \cdot a \ldots (b \; times)$

where a is the base (the number repeatedly multiplied)
b is the exponent or power (indicates how many times to repeatedly multiply the base)

Examples:

1) $4^4 = 4 \cdot 4 \cdot 4 \cdot 4 = 256$
2) $(-4)^4 = -4 \cdot -4 \cdot -4 \cdot -4 = 256$
3) $-4^4 = -(4 \cdot 4 \cdot 4 \cdot 4) = -256$ *note: This problem reads as take the opposite of 4^4.*
4) $-(-4)^4 = -(-4 \cdot -4 \cdot -4 \cdot -4) = -(256) = -256$ *note: This problem reads as take the opposite of $(-4)^4$. Therefore, in following order of operations, you simplify the power first, then take the opposite of that answer.*

In the same previous section, you were given the order of operations to follow when simplifying expressions. The numbers in those problems were whole numbers. Well, the order of operations will still hold true for problems in which the numbers are positive and negative numbers.

RECALL

ORDER OF OPERATIONS

1. Simplify any operations WITHIN grouping symbols such as parenthesis (), brackets [], braces {}, absolute value ||, and above or below fraction bar.
   Note: If the expression contains more than one grouping symbol, then work from the inner most grouping symbols to outer most grouping symbols.
2. Simplify any exponents/powers and radicals.
3. Simplify division and multiplication as they occur from left to right.
4. Simplify subtraction and addition as they occur from left to right.

NOTE: *The order of operations still apply within the grouping symbols.*

PRACTICE PROBLEMS

1) Simplify.

$-3(-4) - 2(5)$	given expression
$12 - 10$	multiply
2	add/subtract

2) Evaluate.

$-10 \div 2(5)$	given expression
$-5(5)$	divide
-25	multiply

Note: In this expression the parenthesis is playing the role of multiplication symbol and not a grouping symbol. Therefore, you can rewrite the expression as $-10 \div 2 \cdot 5$ and continue the steps.

3) Perform the indicated operations.

$-24 \div 3 \cdot 4 - 2^5 \div 2^3 \cdot 7$	given expression
$-24 \div 3 \cdot 4 - 32 \div 8 \cdot 7$	simplify exponents
$-8 \cdot 4 - 32 \div 8 \cdot 7$	divide
$-32 - 32 \div 8 \cdot 7$	multiply
$-32 - 4 \cdot 7$	divide
$-36 - 28$	multiply
-64	add/subtract

4) Simplify.

$5 + 7(9 - 12)$	given expression
$5 + 7(-3)$	simplify within parenthesis
$5 + -21$	multiply
-16	add/subtract

5) Simplify.

$8 - 3[9 - 2(-7 - 5)^2]$	given expression
$8 - 3[9 - 2(-12)^2]$	simplify within parenthesis
$8 - 3[9 - 2(144)]$	simplify exponents
$8 - 3[9 - 288]$	multiply within brackets
$8 - 3[-279]$	add/subtract within brackets
$8 + 837$	multiply
845	add/subtract

6) Simplify

$48 - 4[32 - (5 - 11)^2 + 2]^3 \div 2 \cdot -3$	given expression
$48 - 4[32 - (-6)^2 + 2]^3 \div 2 \cdot -3$	simplify within parenthesis
$48 - 4[32 - 36 + 2]^3 \div 2 \cdot -3$	simplify exponents within brackets
$48 - 4[-2]^3 \div 2 \cdot -3$	add/subtract within brackets
$48 - 4[-8] \div 2 \cdot -3$	simplify exponents
$48 + 32 \div 2 \cdot -3$	multiply
$48 + 16 \cdot -3$	divide
$48 - 48$	multiply
0	add/subtract

EXERCISE 2.3 SET A

Find each answer. Follow the order of operations.

1) $-55 - 60 + 6$

2) $-12 \div 3 \cdot -2$

3) $43 - 3^4 - 2 \cdot -15$

4) $11(-3) + 5(2) - 2(-2)$

5) $-84 \div -12 \cdot 10 \div -2$

6) $30 - 40 \div -8(2)$

7) $4(-3)(-1)^5 (2)^2$

8) $5 \cdot (-12 - 7 + 6 \cdot -3)$

9) $4 - (1 - 3)^3$

10) $-15 - [25 - 3(-8 + 2)]$

11) $-11 - 5 + (13 - 2^2)$

12) $[-18 - (-30)] \div 2 \cdot 4$

13) $122 - [-3 + (4 - 3^2) \cdot -2]$

14) $-(-3)^4 \div 27 - (35 - 32)^2$

15) $2(3 - 6) - 3[15 - (-6) - 9)]$

16) $20 - 3[5^3 - 50 \div -2 \cdot (-3 + 2)]$

17) $-25 + 6 - \{3[5 - (2 - 3)^5 + 4(6 - 4)^3 \div 4]\}$

18) $3 - (-4 - 2) \div 2 \cdot -4 - (8 - 7)^3$

19) $(-2)^3 + \{55 - [3(-12) + (-2 + 8)^2 \div 6] + 1\}$

20) $[(-8)^2 - 5 \cdot 4 \div (10 - 5)] + (-12 + 5)^2$

21) $\dfrac{10 \cdot (-3)^3 - (5 + 15 - 10)^2}{-5(4 - 2)}$

22) $\dfrac{-12 + 5(6 - 3)^2 \div 9}{5 - 5^2 + (5 - 2)^3}$

EXERCISE 2.3 SET B

Find each answer. Follow the order of operations.

1) $-105 + 40 - 95$

2) $48 \div -3 \cdot -2$

3) $100 - 3^4 + 2 \cdot -15$

4) $12(-5) + 5(-5) - 12(-3)$

5) $-72 \div 8 \cdot -14 \div 2$

6) $66 - 80 \div 8(5)$

7) $-(-4)^2(3)^2(-1)^6(2)^2$

8) $-(2)^3 \cdot (24 - 20 + 6 \cdot -3)$

9) $[15 - (-1 - 3)^2]^3$

10) $11 - [60 - 3(-2 - 2)]$

11) $21 - (-5) - (13 - 3^3)$

12) $[(18 - 36) \div 2]^2 \cdot -4$

13) $-52 - [-39 \div (4 + 3^2) \cdot 2]$

14) $-(-6)^3 \div 36 - (11 - 8)^3$

15) $6(2 - 4) - 3[24 - (-2)^4 - (9 - 1^4)]$

16) $50 + 4[3^3 - 50 \div 10 \cdot (1 - 2)]$

17) $-16 - 8\{12 - [8 - (13 - 14)^5 + 4(7 - 5)^3 \div 4]\}$

18) $23 - (13 - 15) \div 2 \cdot 8 - (-2 - 1)^3$

19) $11^2 - \{15 - [4(-12) + (3 + 4)^2 \div 7] - 1^4\}$

20) $[(-9)^2 - 8 \cdot 5 \div (10 - 6)] + (13 - 5)^2$

21) $\dfrac{9^2 - [4 + -8 - (-10)^2]}{-5(4 - 3)}$

22) $\dfrac{-25 + 2(5 - 3)^2 \div -8}{(10 - 17) + (6^2 - 4^2)}$

2.4 FROM WORDS TO ALGEBRA

We will look at problems stated in words and determine how to translate them into algebraic statements. After being able to do this, you will see how algebra can be used to solve problems that you may find difficult to solve with just arithmetic.

In order to translate worded statements into algebraic ones, it is important to learn how key words can be translated. In the phrases and statements in the following list, you could use any letter to represent an unknown number or quantity. However, to keep things simple, we will use the letter x.

Concept	Key Word(s)	Phrase or Statement	Algebraic Translation
Addition:	plus	six *plus* a number	$6 + x$
	add	*add* seven to a number	$x + 7$
	sum	the *sum* of a number and two	$x + 2$
	added to	a number *added* to thirteen	$13 + x$
	more than	five *more than* a number	$x + 5$
	increased by	a quantity *increased by* nine	$x + 9$
	total of	the *total of* forty and a number	$40 + x$
	longer than	ten inches *longer than* a quantity	$x + 10$
Subtraction:	minus	a number *minus* six	$x - 6$
	subtract . . . from	*subtract* three *from* a number	$x - 3$
	difference	the *difference* between x and two	$x - 2$
	fewer than	ten *fewer than* a quantity	$x - 10$
	less than	seven *less than* a number	$x - 7$
	decrease by	nine *decreased by* a number	$9 - x$
	shorter than	five feet *shorter than* a length	$x - 5$
Multiplication:	times	six *times* a number	$6x$
	multiplied by	a number *multiplied by* three	$3x$
	product	the *product* of eight and a number	$8x$
	twice	*twice* a number	$2x$
	doubled	a length is *doubled*	$2x$
	tripled	an amount is *tripled*	$3x$
	of	three-fourths *of* a number	$\frac{3}{4}x$
Division:	divided by	a number *divided by* four	$\frac{x}{4}$
	quotient	the *quotient* of a number and -9	$\frac{x}{-9}$
	ratio	a *ratio* of a number and five	$\frac{x}{5}$
Equality:	is equal to	Twice a number *is equal to* 36.	$2x = 36$
	is	The sum of a number and six *is* nine	$x + 6 = 9$
	result . . . is	The *result* of a number tripled *is* 42.	$3x = 42$
	same result as	Increasing a number by six gives the *same result as* doubling the number	$x + 6 = 2x$
	is the same as	The quotient of a number and five *is the same as* the product of 3 and 2.	$\frac{x}{5} = (3)(2)$

CHAPTER 2   57

Example 1
Write the following phrases as algebraic expressions. Use x for the unknown number.

Word Phrase	Algebraic Expression
a. the sum of seven and a number	a. $7 + x$
b. the product of -5 and a number	b. $-5x$
c. the difference between twice a number and nineteen	c. $2x - 19$
d. five more than the quotient of a number and eight	d. $\dfrac{x}{8} + 5$
e. twice the sum of a number and five	e. $2(x + 5)$

Example 2
Write the following statements as algebraic equations using x for the unknown number or amount.

Word Statement	Algebraic Equation
a. Three less than twice a number is 15.	a. $2x - 3 = 15$
b. The product of a number and five is the same as the number increased by eight.	b. $5x = x + 8$
c. The result of a number tripled is twelve less than the number.	c. $3x = x - 12$
d. The sum of an amount and twice the amount is equal to fifty-seven.	d. $x + 2x = 57$

PROBLEMS

Translate the following into algebraic expressions or equations. Use x for the unknown number.

1) Three times a number decreased by eight.

   Answer: $3x - 8$

2) The product of a number and six is the same as the sum of the number and fifteen.

   Answer: $6x = x + 15$

3) Seven less than twice a number.

   Answer: $2x - 7$

4) The difference between five times a number and six is fifty-four.

   Answer: $5x - 6 = 54$

EXERCISE 2.4 SET A

Write each of the following as an algebraic expression or equation. Use x for the unknown number.

_____ 1) The sum of a number and six

_____ 2) The sum of three and a number

_____ 3) Five fewer than twice a number

_____ 4) Six less than twice a number

_____ 5) The product of a number and six

_____ 6) The product of a number and nine

_____ 7) Four more than three times a number

_____ 8) Two more than five times a number

_____ 9) The quotient of a number and seven

_____ 10) The quotient of a number and four

_____ 11) The sum of twice a number and six is twenty-four.

_____ 12) The difference between a number and two is sixteen.

_____ 13) Twice the sum of a number and six is twenty-four.

_____ 14) Four times a number, increased by seven, is -13.

_____ 15) The product of -7 and a number, decreased by five, is 86.

_____ 16) Ten less than a number tripled gives the same result as the total of the number and six.

_____ 17) The sum of twice a number, three times the number, and four times the number is ninety.

_____ 18) Three times the difference between a number and five is equal to -25 increased by four.

_____ 19) If the sum of a number and three is multiplied by four, the result is forty-eight.

_____ 20) If twice the sum of a number and three is increased by the number, the result is equal to the sum of the number and eighteen.

EXERCISE 2.4 SET B

Write each of the following as an algebraic expression or equation. Use x for the unknown number.

_____ 1) The sum of a number and two

_____ 2) The sum of eight and a number

_____ 3) Seven less than twice a number

_____ 4) Four fewer than twice a number

_____ 5) The product of a number and five

_____ 6) The product of three and a number

_____ 7) Six more than four times a number

_____ 8) One more than three times a number

_____ 9) The quotient of a number and nine

_____ 10) The quotient of a number and two

_____ 11) The sum of twice a number and four is eighteen.

_____ 12) The difference between a number and five is twelve.

_____ 13) Twice the sum of a number and three is thirty-two.

_____ 14) Five times a number, increased by nine, is 39.

_____ 15) The product of −6 and a number, decreased by five, is 67.

_____ 16) Four less than a number tripled gives the same result as the total of the number and twelve.

_____ 17) The sum of twice a number, three times the number, and five times the number is ninety.

_____ 18) Four times the difference between a number and six is equal to −21 increased by five.

_____ 19) If the sum of a number and four is multiplied by three, the result is eighteen.

_____ 20) If twice the sum of a number and five is increased by the number, the result is equal to the sum of the number and sixteen.

3.1 FRACTIONS AND MIXED NUMBERS

CHAPTER 3

He owes being elected to de-nominator.

You have seen numbers such as $\frac{1}{2}, \frac{5}{8}, \frac{1}{4}, \frac{3}{16}, \frac{2}{7}$, etc. These numbers are called **fractions.** They are not whole numbers since they represent a part or a portion of a whole.

For example, what part of the figure below is shaded?

One out of the three equal sections is shaded. So we write $\frac{1}{3}$ (one-third) of the whole figure is shaded.

What part of the figure below is shaded?

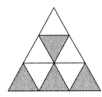

Four out of the nine equal sections are shaded. So we write $\frac{4}{9}$ (four-ninths) of the whole figure is shaded.

Every fraction has three parts. The top number is the numerator, the bottom number is the denominator, and middle line is the fraction line.

---

fraction line → $\frac{5}{16}$ ← **numerator** (tells how many parts you have)

← **denominator** (tells the total number of parts)

PROBLEMS

1) What part of the figure is:

shaded?

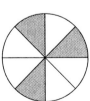

Answer: $\frac{3}{8}$, since 3 out of the 8 sections are shaded.

not shaded?

Answer: $\frac{5}{8}$, since 5 out of the 8 sections are not shaded.

2) If on a 50 question test you get 37 correct, what fraction did you get correct?

Answer: $\frac{37}{50}$

1.) Pages 009–018, 037–042, 045–047, 052–059, 062–077, 080–087, 115–127, 129–132, 141–143, 147–152, 184–187, 199–203, 206–209, from *Math for College Students: Arithmetic with Introductions to Algebra and Geometry,* 5th Edition by Ronal Straszkow. © 1997 by Kendall/Hunt Publishing Company. Used with permission.

PROPER FRACTIONS

All the fractions shown so far have displayed a portion of the whole object or a part of one unit. Each of those fractions has a value that is less than 1. We call such fractions **proper fractions.** Proper fractions are easy to recognize since, besides having a value that is less than 1, their numerators are less than their denominators.

For example:

$\dfrac{3}{5}$ ⟵ numerator is less than the denominator

IMPROPER FRACTIONS

Besides proper fractions that have a value that is less than 1, there are other fractions which have a value that is either equal to 1 or greater than 1. These are called **improper fractions.** Let me show you some examples:

Example 1 In the figure below, four out of the four sections are shaded. $\dfrac{4}{4}$ (four-fourths) is shaded.

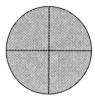

But that represents the whole object. So $\dfrac{4}{4} = 1$.

Thus, $\dfrac{4}{4}$ is a fraction whose value is not *part* of a whole; it *is* the whole. It equals 1.

Likewise $\dfrac{8}{8} = 1$, $\dfrac{10}{10} = 1$, $\dfrac{327}{327} = 1$, etc.

Example 2 In the figures below, we have $\dfrac{4}{4}$ of figure A shaded and $\dfrac{3}{4}$ of figure B shaded. So, a total of $\dfrac{7}{4}$ is shaded.

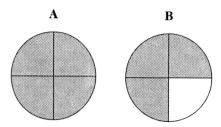

We have one whole object and $\dfrac{3}{4}$ of another that is shaded.

We have 1 and $\dfrac{3}{4}$ (written $1\dfrac{3}{4}$) shaded. So $\dfrac{7}{4} = 1\dfrac{3}{4}$.

Thus, $\dfrac{7}{4}$ is a fraction whose value is greater than 1. Similarly $\dfrac{13}{5}$, $\dfrac{110}{17}$, $\dfrac{9}{2}$, etc., have values that are larger than 1.

Improper fractions such as those are also easily recognized since, besides having a value that is greater than or equal to 1, their numerators are *not* less than their denominators.

For example: $\dfrac{16}{7}$ ⟵ numerators are *not* less ⟶ $\dfrac{14}{14}$
than the denominators.

Numbers that are confused are mixed numbers.

Improper fractions have a value that is greater than or equal to 1. For example, in Section 3.1, we showed that $\frac{4}{4} = 1$ and $\frac{7}{4} = 1\frac{3}{4}$. In fact, every improper fraction can be expressed as either a whole number or as the sum of a whole number and a proper fraction (a **mixed number**).

The fraction line not only separates the numerator and the denominator, it also implies a division—the numerator divided by the denominator.

For that reason some improper fractions give whole number results:

$$\frac{4}{4} = 4\overline{)4} = 1$$

$$\frac{15}{3} = 3\overline{)15} = 5$$

$$\frac{120}{12} = 12\overline{)120} = 10$$

Some improper fractions give mixed number results:

$\frac{7}{4} = 4\overline{)7} = 1 \text{ R}3 = 1\frac{3}{4}$ Since we are dividing by 4, the remainder is 3 out of 4, or $\frac{3}{4}$. So the mixed number is 1 and $\frac{3}{4}$.

$\frac{19}{5} = 5\overline{)19} = 3 \text{ R}4 = 3\frac{4}{5}$ Since we are dividing by 5, the remainder is 4 out of 5, or $\frac{4}{5}$. So the mixed number is 3 and $\frac{4}{5}$.

To simplify an improper fraction all you need to do is to divide the numerator by the denominator and represent any remainder as a proper fraction with the divisor as its denominator. *If you get confused on which number to divide by, you can remember* **D** *for* **d**enominator, **d**own number, *and* **d**ivisor.

PROBLEMS

1) Simplify $\frac{18}{3}$.   *Answer:* 6

$$\frac{18}{3} = 3\overline{)18} = 6$$

2) Simplify $\frac{39}{7}$.   *Answer:* $5\frac{4}{7}$

$$\frac{39}{7} = 7\overline{)39} = 5 \text{ R}4 = 5\frac{4}{7}$$

In working with fractions, you will find it sometimes necessary to change whole or mixed numbers into improper fractions. This is the reverse of what we just did in this section.

For example: $\frac{2}{1} = 2 \div 1 = 2$

Thus, the whole number 2 is the improper fraction $\frac{2}{1}$.

Similarly, any whole number can be converted into an improper fraction by putting the whole number over a denominator of 1.

That is, $7 = \frac{7}{1}$, $25 = \frac{25}{1}$, $156 = \frac{156}{1}$, and so on.

We can also obtain a method for converting mixed numbers into improper fractions by studying how we changed improper fractions into mixed numbers earlier in this section.

$$\text{For example: } \frac{7}{4} = 1\frac{3}{4}$$

$$\frac{7}{4} = 4\overline{)7} = 1 \text{ R}3 = 1\frac{3}{4}$$

with dividend (7), quotient (1), remainder (3), and divisor (4) labeled.

Thus, if you have a mixed number, you have the quotient, the remainder, and the divisor of a division problem. All you need, to find the improper fraction, is the dividend.

But remember, the dividend = divisor × quotient + remainder.

But remember, the dividend = divisor × quotient + remainder

$$1\frac{3}{4} = \frac{4 \times 1 + 3}{4} = \frac{7}{4}$$

If you understand that process for changing a mixed number into a fraction, the short cut below will make it easier.

$$5\frac{3}{4} = \frac{23}{4} \qquad 10\frac{1}{6} = \frac{61}{6} \qquad 122\frac{2}{3} = \frac{368}{3}$$

PROBLEMS

3) Change $6\frac{2}{7}$ to an improper fraction.   Answer: $\frac{44}{7}$

$$6\frac{2}{7} = \frac{7 \times 6 + 2}{7} = \frac{44}{7}$$

4) Change $140\frac{1}{2}$ to an improper fraction.   Answer: $\frac{281}{2}$

$$140\frac{1}{2} = \frac{281}{2}$$

EXERCISE 3.1 SET A

State if the sentence is true or false.

1) A proper fraction always has a value between 0 and 1.

2) An improper fraction always has a value greater than or equal to 1.

3) A mixed number always has a value greater than 1.

4) $\dfrac{0}{5} = 0$

5) $\dfrac{5}{0}$ is undefined.

6) $\dfrac{3}{3}$ is an improper fraction.

7) $\dfrac{21}{22}$ has a value closer to 1 than 0.

8) $4\dfrac{3}{8}$ has a value closer to 5 than 4.

9) $\dfrac{9}{11}$ is a proper fraction and has a value less than one.

10) A mixed number consists of a whole number plus a proper fraction. Therefore, $4\dfrac{5}{8}$ is a mixed number, whereas $4\dfrac{8}{5}$ is not a mixed number.

Use the following set of fractions to categorize them.

$$\left\{\dfrac{1}{2},\ \dfrac{8}{5},\ 4\dfrac{2}{3},\ \dfrac{5}{8},\ \dfrac{8}{8},\ \dfrac{7}{3},\ \dfrac{8}{1},\ 2\dfrac{1}{2}\right\}$$

11) Which numbers are proper fractions?

12) Which numbers are improper fractions?

13) Which numbers are mixed numbers?

Find each fraction.

14) On a true/false test with 100 questions, Portia got 77 of them correct. What fraction of the questions did Portia get correct?

15) In a company with 55 employees, 31 of them are males. What fraction of the employees are males?

16) In a survey of 511 five-year olds, 337 of them said they like ice-cream better than chocolate cake. What fraction of the five-year olds like chocolate cake better than ice-cream?

17) In a football stadium, there are 2000 end-zone seats. On Sunday, one thousand, three hundred three of the end zone seats were filled. What fraction of the end zone seats were not filled?

18) When Brianna cracked open her piggy bank she found 113 one-dollar bills, 51 quarters, 111 dimes, 21 nickels and 543 pennies.
    a) What fractional part of the coins are dimes?
    b) What fractional part of the coins are nickels and pennies?

19) The results of a recent survey on the best vacation spots stated the following:

   143 Bahamas   73 Italy
   151 Aruba   61 Germany
   99 Hawaii   83 France

   a) How many people were surveyed?
   b) What fraction of the people surveyed stated that Aruba was the best vacation spot?
   c) What fraction of the people surveyed stated that the best vacation spot was a European country?
   d) What fraction of the people surveyed stated that the best vacation spot was on an island?

20) A factory that produces tennis rackets found 29 of them to be defective yesterday. If 1000 tennis rackets of them were produced yesterday, what fractional part of them were found to be non-defective?

EXERCISE 3.1 SET B

In problems 1–14, change the fractions to whole or mixed numbers.

_____ 1) $\dfrac{8}{1}$ _____ 2) $\dfrac{7}{1}$

_____ 3) $\dfrac{15}{3}$ _____ 4) $\dfrac{24}{6}$

_____ 5) $\dfrac{17}{3}$ _____ 6) $\dfrac{25}{6}$

_____ 7) $\dfrac{10}{7}$ _____ 8) $\dfrac{11}{8}$

_____ 9) $\dfrac{37}{5}$ _____ 10) $\dfrac{47}{5}$

_____ 11) $\dfrac{125}{16}$ _____ 12) $\dfrac{140}{17}$

_____ 13) $\dfrac{457}{7}$ _____ 14) $\dfrac{396}{7}$

In problems 15–30, change the whole or mixed numbers to improper fractions.

_____ 15) 9   _____ 16) 8

_____ 17) $1\frac{1}{2}$ _____ 18) $1\frac{3}{5}$

_____ 19) $2\frac{5}{8}$ _____ 20) $2\frac{6}{7}$

_____ 21) $9\frac{1}{4}$ _____ 22) $8\frac{1}{3}$

_____ 23) $37\frac{1}{2}$ _____ 24) $52\frac{1}{6}$

_____ 25) $10\frac{5}{6}$ _____ 26) $20\frac{8}{9}$

_____ 27) $124\frac{3}{7}$ _____ 28) $209\frac{2}{3}$

_____ 29) $45\frac{15}{17}$ _____ 30) $19\frac{19}{32}$

3.2 EQUIVALENT FRACTIONS

Oil reduces fraction.

Consider the two shaded blocks below:

 $\frac{3}{4}$ of the block is shaded.

 $\frac{6}{8}$ of the block is shaded.

The shaded sections represent the same amount of each block. So those two fractions, $\frac{3}{4}$ and $\frac{6}{8}$, represent the same amount of shaded area. They have the same value.

$\frac{3}{4}$ and $\frac{6}{8}$ are equivalent. $\frac{3}{4} = \frac{6}{8}$

You can change $\frac{3}{4}$ to $\frac{6}{8}$ by multiplying its numerator and denominator by 2. You can change $\frac{6}{8}$ to $\frac{3}{4}$ by dividing its numerator and denominator by 2. Both procedures yield fractions that have the same value.

RAISING FRACTIONS TO HIGHER TERMS

By multiplying the numerator and denominator of a fraction by the same number, you **raise the fraction to higher terms**, getting an equivalent fraction.

$$\frac{3}{4} \xrightarrow{\times 2} \frac{6}{8} \qquad \frac{1}{3} \xrightarrow{\times 3} \frac{3}{9} \qquad \frac{5}{8} \xrightarrow{\times 10} \frac{50}{80}$$

Example 1

Change $\frac{3}{8}$ to a fraction with a numerator of 12.

$$\frac{3}{8} \xrightarrow{\times 4} \frac{12}{?} = \frac{12}{32}$$

To change the numerator from 3 to 12, you must multiply by 4. So you must also multiply the denominator by 4.

Example 2

Change $\frac{3}{2}$ to a fraction with a denominator of 18.

$$\frac{3}{2} \xrightarrow{\times 9} \frac{?}{18} = \frac{27}{18}$$

To change the denominator from 2 to 18, you must multiply by 9. So you must also multiply the numerator by 9.

PROBLEM

1) $\frac{5}{6} = \frac{35}{?}$

Answer: $\frac{35}{42}$ (Multiply numerator and denominator by 7.)

REDUCING FRACTIONS TO LOWEST TERMS–METHOD 1

To **reduce a fraction to lower terms**, you must find a number other than 1 that divides evenly into both the numerator and denominator of the fraction. To **reduce to lowest terms**, you must continue to divide both the numerator and the denominator until no whole number except 1 divides evenly into each.

Example 3

Reduce $\frac{30}{42}$ to lowest terms.

2 divides evenly into both 30 and 42.

$$\frac{30}{42} \xrightarrow{\div 2} \frac{15}{21}$$

Now, 3 divides evenly into both 15 and 21.

Example 4

Reduce $\frac{140}{910}$ to lowest terms.

$$\frac{30}{42} = \frac{15}{21} \xrightarrow{\div 3} \frac{5}{7}$$

Since numbers that end in zero are divisible by 10, 10 divides evenly into both 140 and 910.

$$\frac{140}{910} \xrightarrow{\div 10} \frac{14}{91}$$

Now, 7 divides evenly into both 14 and 91.

$$\frac{140}{910} = \frac{14}{91} \xrightarrow{\div 7} \frac{2}{13}$$

The key to reducing fractions by this method is to find numbers that divide evenly into both the numerator and the denominator of the fraction. Using the divisibility facts covered in Section 2.9 will be very helpful in this process.

PROBLEMS

2) Reduce $\frac{12}{24}$ to lowest terms. *Answer:* $\frac{1}{2}$

$$\frac{12}{24} \xrightarrow{\div 3} \frac{4}{8} \xrightarrow{\div 2} \frac{2}{4} \xrightarrow{\div 2} \frac{1}{2}$$

3) Reduce $\frac{135}{360}$ to lowest terms. *Answer:* $\frac{3}{8}$

$$\frac{135}{360} \xrightarrow{\div 5} \frac{27}{72} \xrightarrow{\div 3} \frac{9}{24} \xrightarrow{\div 3} \frac{3}{8}$$

REDUCING FRACTIONS TO LOWEST TERMS-METHOD 2

Finding numbers that divide evenly into both the numerator and the denominator of a fraction may be troublesome, even with the use of the divisibility facts learned previously. This second method allows you to work separately with the numerator and denominator, instead of searching for a divisor of both.

Example 5

Reduce $\dfrac{28}{210}$ to lowest terms.

1. Factor the numerator into primes: $28 = 2 \times 2 \times 7$
2. Factor the denominator into primes: $210 = 2 \times 3 \times 5 \times 7$
3. Cancel like factors:

$$\frac{28}{210} = \frac{\cancel{2} \times 2 \times \cancel{7}}{\cancel{2} \times 3 \times 5 \times \cancel{7}}$$

(If the numerator and denominator have the same factor, divide both by that number, giving the answer of 1 $\left(\frac{1}{1}\right)$. This process is called **canceling**.)

4. Multiply the resulting factors: $\dfrac{28}{210} = \dfrac{1 \times 2 \times 1}{1 \times 3 \times 5 \times 1} = \dfrac{2}{15}$

By following the steps shown in the above example, you have a second method for reducing fractions to lowest terms. However, you must know how to factor numbers into primes to use this method. See Section 2–9, if you need to review that factoring process.

Example 6

Reduce $\dfrac{52}{78}$ to lowest terms.

$$\frac{52}{78} = \frac{\cancel{2} \times 2 \times \cancel{13}}{\cancel{2} \times 3 \times \cancel{13}} = \frac{2}{3}$$

Example 7

Reduce $\dfrac{85}{102}$ to lowest terms.

$$\frac{85}{102} = \frac{5 \times \cancel{17}}{2 \times 3 \times \cancel{17}} = \frac{5}{6}$$

PROBLEMS

4) Reduce $\dfrac{60}{105}$ to lowest terms.   Answer: $\dfrac{4}{7}$

$$\frac{60}{105} = \frac{2 \times 2 \times \cancel{3} \times \cancel{5}}{7 \times \cancel{3} \times \cancel{5}} = \frac{4}{7}$$

5) Reduce $\dfrac{99}{154}$ to lowest terms.   Answer: $\dfrac{9}{14}$

$$\frac{99}{154} = \frac{3 \times 3 \times \cancel{11}}{2 \times 7 \times \cancel{11}} = \frac{9}{14}$$

72 CHAPTER 3

We can summarize the methods for **reducing fractions** as follows:

REDUCING FRACTIONS

Method 1—Divide the numerator and denominator by a whole number (larger than 1) that divides evenly into both. Repeat until no more divisors can be found.

Method 2—Factor the numerator and denominator into primes, cancel factors that are the same in both, and multiply the resulting numbers.

Both methods work quite effectively in reducing fractions to lowest terms. Use the method that you find easier to work with.

REDUCING "TROUBLESOME" FRACTIONS

Even with the use of Method 1 and Method 2, a fraction may be difficult to reduce or simply cannot be reduced since it is already in lowest terms. With such fractions a divisor of both the numerator and denominator may not be obvious. When this happens, find the divisors of one of them and see if any of those divisors divide into the other one.

Example 8

Reduce $\frac{87}{203}$ to lowest terms.

It is not obvious if any number divides evenly into both 87 and 203. We know that 3 divides into 87 since the sum of its digits is divisible by 3.

$$87 = 3 \times 29$$

Thus, the only possible divisors that can be used to reduce the fraction is 3 or 29. But 3 does not divide into 203, so 29 is the only possible divisor. 29 divides evenly into 203.

$$\begin{array}{r} 7 \\ 29{\overline{\smash{\big)}\,203}} \\ \underline{203} \end{array}$$

Thus, by dividing the numerator and denominator by 29, we get:

$$\frac{87}{203} \xrightarrow{\div 29} = \frac{3}{7}$$

Example 9

Reduce $\frac{329}{705}$ to lowest terms.

It is not obvious if any number divides evenly into both 329 and 705. However, 5 divides into 705 since it ends in 5. So let's find the divisors of 705.

$$705 = 5 \times 141$$
$$= 5 \times 3 \times 47 \text{ (141 is divisible by 3, sum of its digits is 6.)}$$

Thus, the only numbers that can be used to reduce the fraction are 5, 3, or 47. But 3 and 5 do not divide evenly into 329, so try 47.

$$\frac{329}{705} \xrightarrow[\div 47]{\div 47} \frac{7}{15}$$

Example 10   Reduce $\frac{238}{429}$ to lowest terms.

It is not obvious if any number divides evenly into both 238 and 429. Let's find the divisors of 238 and see if any of them divide into 429.

$$238 = 2 \times 119$$
$$= 2 \times 7 \times 17$$

Thus, the only numbers that can be used to reduce the fraction are 2, 7, or 17. But, none of them divide evenly into 429.

$$429 \div 2 = 214 \text{ R1}, \quad 429 \div 7 = 61 \text{ R2}, \quad 429 \div 17 = 25 \text{ R4}$$

Thus, $\frac{238}{429}$ is in lowest terms and cannot be reduced.

PROBLEMS

6) Reduce $\frac{100}{150}$ to lowest terms.   *Answer:* 2/3

Method 1:

$$\frac{100}{150} \xrightarrow[\div 10]{\div 10} \frac{10}{15} \xrightarrow[\div 5]{\div 5} \frac{2}{3}$$

Method 2:

$$\frac{100}{150} = \frac{\cancel{2} \times 2 \times \cancel{5} \times \cancel{5}}{\cancel{2} \times 3 \times \cancel{5} \times \cancel{5}} = \frac{2}{3}$$

7) Reduce $\frac{63}{143}$ to lowest terms.   *Answer:* 5/11

$$\frac{65}{143} = \frac{5 \times \cancel{13}}{11 \times \cancel{13}} = \frac{5}{11}$$

8) Reduce $\frac{36}{90}$ to lowest terms.

Answer: 2/5

$$\frac{36}{90} \xrightarrow{\div 9}_{\div 9} \frac{4}{10} \xrightarrow{\div 2}_{\div 2} \frac{2}{5}$$

9) Reduce $\frac{182}{231}$ to lowest terms.

Answer: 26/33

The divisors of 231 are 3, 7, and 11. The only divisor that divides evenly into the numerator is 7.

$$\frac{182}{231} \xrightarrow{\div 7}_{\div 7} \frac{26}{33}$$

10) Reduce $\frac{370}{399}$ to lowest terms.

Answer: 370/399

The divisors of 370 are 2, 5, and 37. None of those divide into 399. So, the fraction is in lowest terms.

Name _____ **Date** _____

EXERCISE 3.2

In problems 1–10, raise each fraction to higher terms.

_____ 1) $\frac{1}{2} = \frac{?}{6}$ 　　　　　_____ 2) $\frac{1}{2} = \frac{?}{8}$

_____ 3) $\frac{2}{3} = \frac{?}{9}$ 　　　　　_____ 4) $\frac{3}{5} = \frac{?}{15}$

_____ 5) $\frac{8}{7} = \frac{32}{?}$ 　　　　　_____ 6) $\frac{9}{8} = \frac{36}{?}$

_____ 7) $\frac{6}{5} = \frac{72}{?}$ 　　　　　_____ 8) $\frac{3}{4} = \frac{27}{?}$

_____ 9) $\frac{1}{6} = \frac{?}{180}$ 　　　　　_____ 10) $\frac{1}{6} = \frac{?}{72}$

In problems 11–32, reduce each fraction to lowest terms.

_____ 11) $\frac{3}{15}$ 　　　　　_____ 12) $\frac{4}{12}$

_____ 13) $\frac{10}{18}$ 　　　　　_____ 14) $\frac{6}{9}$

_____ 15) $\dfrac{21}{33}$ _____ 16) $\dfrac{27}{39}$

_____ 17) $\dfrac{80}{120}$ _____ 18) $\dfrac{90}{240}$

_____ 19) $\dfrac{75}{180}$ _____ 20) $\dfrac{132}{180}$

_____ 21) $\dfrac{78}{441}$ _____ 22) $\dfrac{84}{351}$

_____ 23) $\dfrac{108}{144}$ _____ 24) $\dfrac{195}{255}$

_____ 25) $\dfrac{34}{85}$ _____ 26) $\dfrac{51}{68}$

_____ 27) $\dfrac{52}{91}$ _____ 28) $\dfrac{78}{91}$

_____ 29) $\dfrac{138}{161}$ _____ 30) $\dfrac{205}{246}$

_____ 31) $\dfrac{110}{399}$ _____ 32) $\dfrac{165}{182}$

3.3 MULTIPLY AND DIVIDE WITH FRACTIONS AND MIXED NUMBERS

The easiest operation with fractions is multiplication, since to multiply fractions you just have to multiply their numerators and multiply their denominators to get the numerator and the denominator of the answer.

For example: $\dfrac{1}{2} \times \dfrac{3}{5} = \dfrac{1 \times 3}{2 \times 5} = \dfrac{3}{10}$

$$\dfrac{4}{7} \times \dfrac{1}{5} \times \dfrac{6}{13} = \dfrac{4 \times 1 \times 6}{7 \times 5 \times 13} = \dfrac{24}{455}$$

The difficulty, however, with working with fractions is that answers should be reduced to lowest terms.

For example: $\dfrac{6}{15} \times \dfrac{42}{63} = \dfrac{6 \times 42}{15 \times 63} = \dfrac{252}{945}$

We must now reduce that answer:

$$\dfrac{252}{945} \xrightarrow{\div 3} \dfrac{84}{315} \xrightarrow{\div 3} \dfrac{28}{105} \xrightarrow{\div 7} \dfrac{4}{15}$$

The reducing process in that example was quite involved since the numerator and the denominator of the fraction were large numbers. However, if we reduced the fractions before we multiplied, the solution would be a little easier.

Consider $\dfrac{6}{15} \times \dfrac{42}{63}$ again: $\dfrac{6}{15}$ reduces to $\dfrac{2}{5}$ and $\dfrac{42}{63}$ reduces to $\dfrac{2}{3}$.
($\div$ by 3) ($\div$ by 21)

Thus, $\dfrac{6}{15} \times \dfrac{42}{63} = \dfrac{2}{5} \times \dfrac{2}{3} = \dfrac{4}{15}$.

NOTE: To eliminate some of the writing involved in doing a problem such as that, we would write the problem as shown below:

$$\dfrac{\cancel{6}^{2}}{\cancel{15}_{5}} \times \dfrac{\cancel{42}^{2}}{\cancel{63}_{3}} = \dfrac{2 \times 2}{5 \times 3} = \dfrac{4}{15}$$

Notice that the canceling of numbers is similar to what was previously done.

There is another short cut that can be used when multiplying fractions. Let me show you by another example. Consider:

$$\dfrac{3}{70} \times \dfrac{14}{25} = \dfrac{3 \times 14}{70 \times 25} = \dfrac{42}{1750}$$

We must now reduce that answer:

$$\dfrac{42}{1750} \xrightarrow{\div 2} \dfrac{21}{875} \xrightarrow{\div 7} \dfrac{3}{125}$$

Notice that the original fractions $\frac{3}{70}$ and $\frac{14}{25}$ are not reducible. However, the product $\frac{42}{1750}$ is reducible to lower terms. It would be nice if we could reduce before we multiplied, as we did in the previous example. What we could have done is reduce the numerator of the 2nd fraction with the denominator of the 1st fraction.

$$\frac{3}{\cancel{70}_{5}} \times \frac{\cancel{14}^{1}}{25} = \frac{3 \times 1}{5 \times 25} = \frac{3}{125}$$

This process is called cross canceling and can *only* be used when multiplying fractions. The process requires that you divide the same number evenly into any numerator and any denominator. You can do this canceling process in any order, and you should continue until all possible combinations are canceled.

$$2 \times \frac{18}{10} \times \frac{15}{26} = ?$$

$$= \frac{2}{1} \times \frac{\cancel{18}^{9}}{\cancel{10}_{5}} \times \frac{15}{26} \qquad \text{Change 2 to the fraction } \tfrac{2}{1} \text{ and reduce } \tfrac{18}{10} \text{ to } \tfrac{9}{5}.$$

$$= \frac{\cancel{2}^{1}}{1} \times \frac{9}{\cancel{5}_{1}} \times \frac{\cancel{15}^{3}}{\cancel{26}_{13}} \qquad \text{Cross cancel the 2 and the 26; cross cancel the 5 and the 15.}$$

$$= \frac{27}{13} = 2\frac{1}{13} \qquad \text{Multiply the remaining fractions and change the improper fraction to a mixed number.}$$

PROBLEMS

1) $\dfrac{3}{4} \times \dfrac{20}{7} = ?$ Answer: $2\dfrac{1}{7}$

$$\frac{3}{\cancel{4}_{1}} \times \frac{\cancel{20}^{5}}{7} = \frac{15}{7} = 2\frac{1}{7}$$

2) $\dfrac{12}{15} \times \dfrac{8}{25} \times \dfrac{75}{100} = ?$ Answer: $\dfrac{24}{125}$

$$\frac{\cancel{12}^{4}}{\cancel{15}_{5}} \times \frac{8}{25} \times \frac{\cancel{75}^{3}}{\cancel{100}_{4}} = \frac{4}{5} \times \frac{\cancel{8}^{2}}{25} \times \frac{3}{\cancel{4}_{1}} = \frac{24}{125}$$

DIVIDING FRACTIONS

An understanding of how to multiply fractions is essential in this section. You will learn how division of fractions can be changed into the multiplication of fractions. Look at the problem $12 \div 3$. We can consider dividing a number by 3 the same as taking $\frac{1}{3}$ of it. That is, multiplying it by $\frac{1}{3}$.

$$12 \div 3 = 12 \times \frac{1}{3} = 4$$

Cancel:
A good car salesman can sell.

Also in a problem such as $\frac{3}{5} \div 2$, we can consider dividing a number by 2 the same as taking a half of it.

$$\frac{3}{5} \div 2 = \frac{3}{5} \times \frac{1}{2} = \frac{3}{10}$$

Let's look closely at those two problems. We know that $3 = \frac{3}{1}$ and $2 = \frac{2}{1}$, so those two problems become:

$$12 \div 3 = 12 \div \frac{3}{1} = 12 \times \frac{1}{3} \qquad \frac{3}{5} \div 2 = \frac{3}{5} \div \frac{2}{1} = \frac{3}{5} \times \frac{1}{2}$$

Those two examples point out a method for dividing fractions.

DIVIDING FRACTIONS

Invert the divisor (find its **reciprocal**) and change the division into multiplication.

$$12 \div \underset{(\div \text{ changed to } \times)}{\frac{3}{1}} = 12 \times \overset{(\text{reciprocal})}{\frac{1}{3}} \qquad \frac{3}{5} \div \underset{(\div \text{ changed to } \times)}{\frac{2}{1}} = \frac{3}{5} \times \overset{(\text{reciprocal})}{\frac{1}{2}}$$

Using that method, problems that involve dividing fractions become multiplication problems with the original divisor inverted. Let's try some examples.

Example 1

$$\frac{2}{3} \div \frac{5}{7} = ?$$
$$= \frac{2}{3} \times \frac{7}{5} \qquad \text{Invert the divisor and multiply.}$$
$$= \frac{14}{15}$$

Example 2

$$\frac{10}{15} \div \frac{7}{27} = ?$$

$$= \frac{10}{15} \div \frac{7}{27} \quad \text{Invert the divisor and multiply.}$$

$$= \frac{\overset{2}{\cancel{10}}}{\underset{3}{\cancel{15}}} \times \frac{27}{7} \quad \text{Reduce.}$$

$$= \frac{2}{\underset{1}{\cancel{3}}} \times \frac{\overset{9}{\cancel{27}}}{7} \quad \text{Cross cancel.}$$

$$= \frac{18}{7} = 2\frac{4}{7} \quad \text{Multiply and change the improper fraction to a mixed number.}$$

Example 3

$$35 \div \frac{28}{6} = ?$$

$$= 35 \times \frac{6}{28} \quad \text{Invert the divisor and multiply.}$$

$$= \frac{35}{1} \times \frac{3}{14} \quad \text{Change 35 to } \frac{35}{1} \text{ and reduce } \frac{6}{28} \text{ to } \frac{3}{14}.$$

$$= \frac{\overset{5}{\cancel{35}}}{1} \times \frac{3}{\underset{2}{\cancel{14}}} \quad \text{Cross cancel.}$$

$$= \frac{15}{2} = 7\frac{1}{2} \quad \text{Multiply and change the improper fraction to a mixed number.}$$

We can summarize **division of fractions** as follows:

1. Invert the divisor and change to multiplication.
2. Reduce and cross cancel, if possible.
3. Multiply the resulting fractions.
4. Change any improper fraction into a mixed number.

PROBLEM

1) $\frac{10}{3} \div 45 = ?$ Answer: $\frac{2}{27}$

$$\frac{10}{3} \div \frac{45}{1} = \frac{\overset{2}{\cancel{10}}}{3} \times \frac{1}{\underset{9}{\cancel{45}}} = \frac{2}{27}$$

MULTIPLYING AND DIVIDING MIXED NUMBERS

You learned how to multiply and divide fractions. If you look over the examples and problems in those sections, you will notice that you operated with both proper and improper fractions. The methods shown worked for both types of fractions. Therefore, the logical way to multiply and divide mixed numbers is to first change them into improper fractions and then use those same methods. Let me show you what I mean.

MULTIPLYING MIXED NUMBERS

Example 1 $11 \times 5\frac{2}{9} = ?$

$= \frac{11}{1} \times \frac{47}{9}$ Change to improper fractions.

$\left(11 = \frac{11}{1} \text{ and } 5\frac{2}{9} = \frac{47}{9}\right)$

$= \frac{517}{9} = 57\frac{4}{9}$ Multiply across and simplify.

Example 2 $3\frac{3}{4} \times 5\frac{1}{3} = ?$

$= \frac{15}{4} \times \frac{16}{3}$ Change the mixed numbers to improper fractions.

$\left(3\frac{3}{4} = \frac{15}{4} \text{ and } 5\frac{1}{3} = \frac{16}{3}\right)$

$= \frac{\cancel{15}^{5}}{\cancel{4}_{1}} \times \frac{\cancel{16}^{4}}{\cancel{3}_{1}}$ Cross cancel the 4 and the 16; cross cancel the 15 and the 3.

$= \frac{20}{1} = 20$ Multiply across and simplify.

PROBLEMS

1) $7\frac{2}{3} \times 10\frac{1}{2} = ?$ *Answer:* $80\frac{1}{2}$

$7\frac{2}{3} \times 10\frac{1}{2} = \frac{23}{\cancel{3}_{1}} \times \frac{\cancel{21}^{7}}{2} = \frac{161}{2} = 80\frac{1}{2}$

2) $8 \times 2\frac{5}{16} = ?$ *Answer:* $18\frac{1}{2}$

$8 \times 2\frac{5}{16} = \frac{\cancel{8}^{1}}{1} \times \frac{37}{\cancel{16}_{2}} = \frac{37}{2} = 18\frac{1}{2}$

DIVIDING MIXED NUMBERS

When dividing mixed numbers, you will again change the mixed numbers into improper fractions. Remember, however, that when you divide fractions you must invert the divisor and change the division to multiplication.

Example 3 $\quad 10\dfrac{2}{15} \div 6\dfrac{2}{5} = ?$

$\quad = \dfrac{152}{15} \div \dfrac{32}{5}$ \qquad Change the mixed numbers to improper fractions.

$\quad = \dfrac{\cancel{152}^{19}}{\cancel{15}_{3}} \times \dfrac{\cancel{5}^{1}}{\cancel{32}_{4}}$ \qquad Invert the divisor; change to multiplication; cross cancel the 15 and the 5, and the 152 and the 32.

$\quad = \dfrac{19}{12} = 1\dfrac{7}{12}$ \qquad Multiply across and simplify.

Example 4 $\quad 9\dfrac{5}{6} \div 12 = ?$

$\quad = \dfrac{59}{6} \div \dfrac{12}{1}$ \qquad Change the mixed numbers to improper fractions.

$\quad = \dfrac{59}{6} \times \dfrac{1}{12}$ \qquad Invert the divisor and multiply.

$\quad = \dfrac{59}{72}$

PROBLEMS

2) $8\dfrac{1}{3} \div 9\dfrac{4}{9} = ?$ \qquad\qquad Answer: $\dfrac{15}{17}$

$$8\dfrac{1}{3} \div 9\dfrac{4}{9} = \dfrac{25}{3} \div \dfrac{85}{9}$$
$$= \dfrac{\cancel{25}^{5}}{\cancel{3}_{1}} \times \dfrac{\cancel{9}^{3}}{\cancel{85}_{17}} = \dfrac{15}{17}$$

3) $76 \div 2\dfrac{5}{8} = ?$ \qquad\qquad Answer: $28\dfrac{20}{21}$

$$76 \div 2\dfrac{5}{8} = \dfrac{76}{1} \div \dfrac{21}{8}$$
$$= \dfrac{76}{1} \times \dfrac{8}{21} = \dfrac{608}{21} = 28\dfrac{20}{21}$$

Name _____   Date _____

EXERCISE 3.3

Find each answer. Reduce to lowest terms.

1) $\dfrac{2}{3} \cdot \dfrac{5}{9}$

2) $\dfrac{1}{4} \cdot \dfrac{3}{5} \cdot \dfrac{9}{2}$

3) $\dfrac{10}{18} \cdot \dfrac{2}{5}$

4) $\dfrac{6}{7} \cdot \dfrac{1}{3}$

5) $\dfrac{24}{30} \cdot \dfrac{12}{18}$

6) $\dfrac{6}{15} \cdot \dfrac{20}{30} \cdot \dfrac{5}{16}$

7) $\dfrac{3}{8} \cdot 32$

8) $18 \cdot \dfrac{3}{4}$

9) $\dfrac{2}{5} \cdot 45 \cdot \dfrac{3}{9}$

10) $\dfrac{5}{11} \cdot \dfrac{1}{2} \cdot 0$

11) $\dfrac{3}{8} \div \dfrac{3}{4}$

12) $\dfrac{14}{20} \div \dfrac{2}{5}$

13) $\dfrac{48}{50} \div \dfrac{16}{30}$

14) $\dfrac{64}{35} \div \dfrac{8}{7}$

15) $\dfrac{45}{55} \div 5$

16) $20 \div \dfrac{4}{7}$

17) $0 \div \dfrac{3}{5}$

18) $\dfrac{14}{15} \div 0$

19) $\dfrac{4}{9} \cdot \dfrac{1}{5} \div \dfrac{8}{18}$

20) $\dfrac{12}{34} \div \dfrac{1}{2} \cdot \dfrac{17}{20}$

21) $\dfrac{15}{21} \div \left(\dfrac{2}{3} \cdot \dfrac{1}{3}\right)$

22) $100 \div \dfrac{2}{5} \cdot \dfrac{1}{50}$

23) $5\dfrac{1}{2} \cdot 3\dfrac{1}{2}$

24) $9\dfrac{3}{4} \cdot 12$

25) $\dfrac{4}{5} \cdot 1\dfrac{1}{3}$

26) $2\dfrac{2}{7} \cdot \dfrac{14}{15} \cdot 10$

27) $8\dfrac{1}{4} \div 2\dfrac{1}{3}$

28) $10\dfrac{5}{8} \div 6\dfrac{1}{8}$

29) $9\dfrac{1}{2} \div 2$

30) $39 \div 2\dfrac{1}{6}$

31) $4\dfrac{2}{3} \div \dfrac{7}{6}$

32) $\dfrac{12}{30} \div 1\dfrac{1}{5}$

33) $3\dfrac{5}{8} \cdot 4\dfrac{1}{4} \div 1\dfrac{1}{2}$

34) $5 \div \dfrac{20}{25} \cdot 2\dfrac{1}{4}$

35) $6\dfrac{2}{7} \cdot \left(2\dfrac{1}{6} \div \dfrac{11}{16}\right)$

36) $11 \div \left(\dfrac{0}{4} \cdot 7\dfrac{1}{7}\right)$

3.4 ADD AND SUBTRACT WITH FRACTIONS AND MIXED NUMBERS

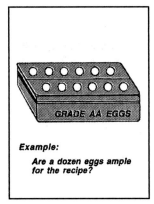

Example:
Are a dozen eggs ample for the recipe?

Add and Subtract with Like Fractions.

The next step in developing your understanding of fractions is learning how to add and subtract fractions. We will start by adding and subtracting **like fractions**, that is, fractions with the same denominators.

For example:

$$\frac{2}{5} + \frac{1}{5} = \frac{3}{5}$$

$$\frac{6}{8} - \frac{1}{8} = \frac{5}{8}$$

Adding and subtracting like fractions is a straightforward process. If you are adding, the numerator of the answer is the sum of the numerators. If you are subtracting, the numerator is the difference between the numerators. In either case, the denominator of the answer is the same as the like denominators in the problem. Notice, we do not add or subtract the denominators. As with all computations with fractions, you should reduce your final answers.

> When adding or subtracting like fractions, leave the denominator the same as in the problem and add or subtract the numerators.

Example 1 $\quad \frac{2}{7} + \frac{3}{7} + \frac{1}{7} = ?$

$\qquad = \frac{2 + 3 + 1}{7} \quad$ Add numerators.

$\qquad = \frac{6}{7} \quad$ Leave the denominator the same.

Example 2 $\quad \frac{5}{6} - \frac{2}{6} = ?$

$\qquad = \frac{5 - 2}{6} \quad$ Subtract the numerators.

$\qquad = \frac{3}{6} \quad$ Leave the denominator the same.

$\qquad = \frac{1}{2} \quad$ Reduce final answer.

Example 3 $\frac{5}{8} + \frac{2}{8} + \frac{3}{8} = ?$

$= \frac{5 + 2 + 3}{8}$

$= \frac{10}{8}$

$= \frac{5}{4}$ Reduce the answer.

$= 1\frac{1}{4}$ Change the improper fraction to a mixed number.

Example 4 $\frac{63}{25} - \frac{11}{25} + \frac{8}{25} = ?$

$= \frac{63 - 11 + 8}{25}$

$= \frac{60}{25}$

$= \frac{12}{5}$ Reduce the answer.

$= 2\frac{2}{5}$ Change the improper fraction to a mixed number.

PROBLEMS

1) $\frac{7}{18} + \frac{3}{18} = ?$ *Answer:* $\frac{5}{9}$

$\frac{7}{18} + \frac{3}{18} = \frac{10}{18} = \frac{5}{9}$

2) $\frac{11}{30} - \frac{5}{30} = ?$ *Answer:* $\frac{1}{5}$

$\frac{11}{30} - \frac{5}{30} = \frac{6}{30} = \frac{1}{5}$

3) $\frac{2}{9} + \frac{5}{9} + \frac{7}{9} - \frac{1}{9} = ?$ *Answer:* $1\frac{4}{9}$

$\frac{2}{9} + \frac{5}{9} + \frac{7}{9} - \frac{1}{9}$

$= \frac{2 + 5 + 7 - 1}{9}$

$= \frac{13}{9}$

$= 1\frac{4}{9}$

ADD AND SUBTRACT WITH UNLIKE FRACTIONS

The fractions in the last section were like fractions since the fractions in each problem had the same denominators. We added and subtracted them very easily. In fact, in order to add and subtract any fractions, they *must* be like fractions. In the problem

$$\frac{3}{4} + \frac{1}{5} = ?$$

we must change each fraction to equivalent fractions that have the same denominators before we can add them. *The denominator that we change each fraction into should be the smallest number that is divisible by each denominator. We call it the least common denominator (LCD).*

In the above problem, the LCD is 20, since 20 is the smallest number that is divisible by both 4 and 5. That means we have to change $\frac{3}{4}$ and $\frac{1}{5}$ into 20ths before we can add them.

$$\frac{3}{4} = \frac{?}{20} \qquad \frac{3}{4} \xrightarrow{\times 5} \frac{15}{20}$$

$$\frac{1}{5} = \frac{?}{20} \qquad +\frac{1}{5} \xrightarrow{\times 4} \frac{4}{20}$$

$$= \frac{19}{20}$$

Now add the like fractions.

Quite frequently you can determine the LCD by inspection. That is, you look at the denominators in an addition or subtraction problem and think, "What is the smallest number that can be evenly divided by each denominator?"

For example, in these problems:

$$\frac{2}{3} - \frac{1}{2} = ?$$ The LCD is 6, since 6 is the smallest number divisible by 3 and 2.

$$\frac{1}{2} + \frac{2}{5} + \frac{9}{10} = ?$$ The LCD is 10, since 10 is the smallest number divisible by 2, 5, and 10.

$$\frac{3}{4} + \frac{5}{6} + \frac{3}{2} - \frac{2}{3} = ?$$ The LCD is 12, since 12 is the smallest number divisible by 4, 6, 2, and 3.

The method, then, for **adding and subtracting fractions** is as follows:

1. Find the least common denominator.
2. Change each fraction to a fraction with the LCD as its denominator.
3. Add or subtract those like fractions.
4. Reduce your answer and change any improper fraction into a mixed number.

Let's use those steps in doing the three previous examples.

Example 1 $\dfrac{2}{3} - \dfrac{1}{2} = ?$

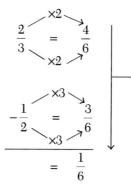

The LCD is 6.

Change each fraction to a fraction with a denominator of 6.

Subtract the like fractions.

Example 2 $\dfrac{1}{2} + \dfrac{2}{5} + \dfrac{9}{10} = ?$

$$\begin{aligned}\dfrac{1}{2} &= \dfrac{5}{10} \\ +\dfrac{2}{5} &= \dfrac{4}{10} \\ +\dfrac{9}{10} &= \dfrac{9}{10} \\ &= \dfrac{18}{10} = \dfrac{9}{5} = 1\dfrac{4}{5}\end{aligned}$$

The LCD is 10.

Change each fraction to a fraction with a denominator of 10.

Add the like fractions.

Reduce the answer and change it to a mixed number.

Example 3 $\dfrac{3}{4} + \dfrac{5}{6} + \dfrac{3}{2} - \dfrac{2}{3} = ?$ The LCD is 12.

$$\begin{aligned}\dfrac{3}{4} &= \dfrac{9}{12} \\ +\dfrac{5}{6} &= \dfrac{10}{12} \\ +\dfrac{3}{2} &= \dfrac{18}{12} \\ -\dfrac{2}{3} &= \dfrac{8}{12} \\ &= \dfrac{29}{12} = 2\dfrac{5}{12}\end{aligned}$$

Change each fraction to a fraction with a denominator of 12.

Add and subtract the like fractions.

Change the improper fraction to a mixed number.

ADD AND SUBTRACT WITH MIXED NUMBERS

Method 1–Change to Improper Fractions

An effective way to add and subtract mixed numbers is to first change them into improper fractions as you did in previous sections. Let me show you how this is done.

Example 1 $\quad 3\frac{2}{7} + 1\frac{8}{21} = ?$

$$3\frac{2}{7} = \frac{23}{7} = \frac{69}{21}$$
$$+1\frac{8}{21} = \frac{29}{21} = \frac{29}{21}$$
$$= \frac{98}{21} = 4\frac{14}{21} = 4\frac{2}{3}$$

Change the mixed numbers to improper fractions.

The LCD is 21. Change each fraction to a denominator of 21.

Add the like fractions, simplify, and reduce.

Example 2 $\quad 40 - 36\frac{3}{5} = ?$

$$40 = \frac{40}{1} = \frac{200}{5}$$
$$-36\frac{3}{5} = \frac{183}{5} = \frac{183}{5}$$
$$= \frac{17}{5} = 3\frac{2}{5}$$

Change the mixed numbers to improper fractions.

The LCD is 5. Change each fraction to a denominator of 5.

Subtract the like fractions and simplify.

PROBLEMS

1) $75\frac{1}{2} - 56\frac{4}{5} = ?$ $\qquad$ *Answer:* $18\frac{7}{10}$

$$75\frac{1}{2} = \frac{151}{2} = \frac{755}{10}$$
$$-56\frac{4}{5} = \frac{284}{5} = \frac{568}{10}$$
$$= \frac{187}{10} = 18\frac{7}{10}$$

2) $2\frac{1}{12} + 3\frac{2}{3} = ?$ $\qquad$ *Answer:* $5\frac{3}{4}$

$$2\frac{1}{12} = \frac{25}{12} = \frac{25}{12}$$
$$+3\frac{2}{3} = \frac{11}{3} = \frac{44}{12}$$
$$= \frac{69}{12} = 5\frac{9}{12} = 5\frac{3}{4}$$

Method 2–Leave as Mixed Numbers

Because the process described in Method 1 may cause you to work with large numbers, it may be easier to operate separately with the fractional and whole number parts when adding or subtracting mixed numbers.

Example 3

$$182\frac{2}{3} + 48\frac{3}{4} = ?$$

$$182\frac{2}{3} = 182\frac{8}{12}$$ Change fractions to LCD of 12.

$$48\frac{3}{4} = 48\frac{9}{12}$$ Add fractional and whole number parts.

$$= 230\frac{17}{12}$$

$$= 230 + 1\frac{5}{12} = 231\frac{5}{12}$$ Change the fraction to a mixed number and add.

Example 4

$$8 - 3\frac{4}{5} = ?$$

Since 8 has no fractional part, you borrow 1 from 8 getting 7. Since the fraction being subtracted has a denominator of 5, the 1 that is borrowed is written as $\frac{5}{5}$.

$$8 = 7\frac{5}{5}$$ Borrow 1 from 8, $1 = \frac{5}{5}$.

$$-3\frac{4}{5} = -3\frac{4}{5}$$ Subtract fractional and whole number parts.

$$= 4\frac{1}{5}$$

Example 5

$$75\frac{3}{8} - 17\frac{3}{4} = ?$$

Since the fractional part being subtracted $\left(\frac{3}{4}\right)$ is larger than $\frac{3}{8}$, you borrow 1 from 75 getting 74. The 1 that is borrowed is written as $\frac{8}{8}$ and added to the $\frac{3}{8}$.

$$75\frac{3}{8} = 74\frac{8}{8} + \frac{3}{8} = 74\frac{11}{8}$$ Borrow 1 from 74, $1 = \frac{8}{8}$.

$$-17\frac{3}{4} = -17\frac{6}{8} = -17\frac{6}{8}$$ Change $\frac{3}{4}$ to LCD, $\frac{3}{4} = \frac{6}{8}$.

$$= 57\frac{5}{8}$$ Subtract both parts.

PROBLEM

3) $26\frac{4}{9} - -19\frac{3}{5} = ?$ Answer: $6\frac{38}{45}$

$$26\frac{4}{9} = 26\frac{20}{45} = 25\frac{45}{45} + \frac{20}{45} = 25\frac{65}{45}$$

$$-19\frac{3}{5} = -19\frac{27}{45} = -19\frac{27}{45} \qquad = -19\frac{27}{45}$$

$$= 6\frac{38}{45}$$

COMPLEX FRACTIONS

This section introduces complex fractions. A **complex fraction** is a fraction that contains more than one fraction line. The objective of this section is to show you how to express a complex fraction as a single fraction.

Example 1

$$\frac{\frac{2}{3}}{\frac{3}{4}} = ?$$

The middle fraction line indicates the fraction above that line is to be divided by the fraction below the line.

$$\frac{\frac{2}{3}}{\frac{3}{4}} = \frac{2}{3} \div \frac{3}{4}$$

$$= \frac{2}{3} \times \frac{4}{3} = \frac{8}{9} \qquad \text{Invert divisor and multiply.}$$

The procedure for evaluating a complex fraction is to determine the answer for the fractions above the middle fraction line and divide it by the results of the fractions below the middle line.

Example 2

$$\frac{\frac{5}{6} + \frac{2}{6}}{4 - \frac{2}{3}} = ?$$

Add the fractions above the line.

$$\frac{5}{6} + \frac{2}{6} = \frac{7}{6}$$

Subtract the numbers below the line.

$$4 - \frac{2}{3} = \frac{4}{1} - \frac{2}{3} = \frac{12}{3} - \frac{2}{3} = \frac{10}{3}$$

Thus, $\dfrac{\frac{5}{6} + \frac{2}{6}}{4 - \frac{2}{3}} = \dfrac{\frac{7}{6}}{\frac{10}{3}} = \dfrac{7}{6} \div \dfrac{10}{3}$ 　　Divide the two results.

$$= \frac{7}{\cancel{6}_2} \times \frac{\cancel{3}^1}{10} = \frac{7}{20} \qquad \text{Invert divisor and multiply.}$$

Example 3

$$\frac{2\frac{2}{3} + 1\frac{3}{8}}{\frac{7}{12} + \frac{11}{30}} = ?$$

$$2\frac{2}{3} = \frac{8}{3} = \frac{64}{24}$$
$$+1\frac{3}{8} = \frac{11}{8} = \frac{33}{24}$$
$$= \frac{97}{24}$$

Add the fractions above the fraction line. (Change mixed numbers to improper fractions and use an LCD of 24.)

$$\frac{7}{12} = \frac{35}{60}$$
$$+\frac{11}{30} = \frac{22}{60}$$
$$= \frac{57}{60} = \frac{19}{20}$$

Add the fractions below the fraction line. (Change to an LCD of 60.)

Reduce the answer (by dividing by 3).

Thus, $\dfrac{2\frac{2}{3} + 1\frac{3}{8}}{\frac{7}{12} + \frac{11}{30}} = \dfrac{\frac{97}{24}}{\frac{19}{20}} = \dfrac{97}{24} \div \dfrac{19}{20}$

Now divide those results.

$$= \frac{97}{\cancel{24}_6} \times \frac{\cancel{20}^5}{19}$$

Invert divisor and multiply.

$$= \frac{485}{114} = 4\frac{29}{114}$$

PROBLEMS

1) $\dfrac{\frac{4}{5}}{3\frac{1}{2}} = ?$ Answer: $\dfrac{8}{35}$

$$\frac{\frac{4}{5}}{\frac{7}{2}} = \frac{4}{5} \div \frac{7}{2} = \frac{4}{5} \times \frac{2}{7} = \frac{8}{35}$$

2) $\dfrac{\frac{2}{3} - \frac{5}{12}}{5 + \frac{1}{2}} = ?$ Answer: $\dfrac{1}{22}$

$$\frac{\frac{8}{12} - \frac{5}{12}}{\frac{5}{1} + \frac{1}{2}} = \frac{\frac{3}{12}}{\frac{10}{2} + \frac{1}{2}} = \frac{\frac{1}{4}}{\frac{11}{2}}$$

$$= \frac{1}{4} \div \frac{11}{2} = \frac{1}{\cancel{4}_2} \times \frac{\cancel{2}^1}{11} = \frac{1}{22}$$

Name _____   Date _____

EXERCISE 3.4

Find each answer. Reduce to lowest terms.

1) $\dfrac{5}{12} + \dfrac{1}{12} + \dfrac{5}{12}$

2) $\dfrac{7}{20} + \dfrac{9}{20} + \dfrac{11}{20}$

3) $\dfrac{14}{25} - \dfrac{8}{25}$

4) $\dfrac{32}{55} - \dfrac{10}{55} - \dfrac{7}{55}$

5) $\dfrac{3}{8} + \dfrac{3}{4}$

6) $\dfrac{1}{3} + \dfrac{2}{5} + \dfrac{5}{6}$

7) $\dfrac{11}{18} + \dfrac{2}{9} + \dfrac{1}{3}$

8) $\dfrac{4}{7} + \dfrac{1}{6} + \dfrac{3}{2}$

9) $\dfrac{16}{21} - \dfrac{2}{3}$

10) $\dfrac{29}{30} - \dfrac{3}{5}$

11) $\dfrac{4}{5} - \dfrac{2}{3}$

12) $\dfrac{21}{4} - \dfrac{10}{3}$

13) $5\dfrac{2}{7} + 1\dfrac{4}{7}$

14) $1\dfrac{3}{8} + 2\dfrac{5}{8} + 3\dfrac{7}{8}$

15) $13\dfrac{7}{10} - 5\dfrac{3}{10}$

16) $8\dfrac{3}{5} - 4$

17) $11\dfrac{4}{5} + 2\dfrac{1}{10}$

18) $4\dfrac{3}{10} + 3\dfrac{7}{20} + 5\dfrac{11}{30}$

19) $9\dfrac{1}{2} + 21 + \dfrac{4}{15}$

20) $32 + 1\dfrac{1}{20} + 2\dfrac{2}{5}$

21) $10\dfrac{5}{8} - 6\dfrac{3}{4}$

22) $8\dfrac{7}{12} - 4\dfrac{2}{5}$

23) $3\dfrac{12}{21} - 2\dfrac{5}{7}$

24) $9\dfrac{10}{25} - \dfrac{2}{5}$

25) $35 - 15\dfrac{8}{9}$

26) $15 - \dfrac{5}{11}$

27) $6\dfrac{1}{2} - 3\dfrac{2}{3}$

28) $4\dfrac{3}{8} - 1\dfrac{4}{7}$

29) $12\dfrac{1}{6} + 3\dfrac{1}{3} - 8\dfrac{3}{4}$

30) $33\dfrac{7}{8} - 21\dfrac{3}{4} + 1\dfrac{1}{16}$

31) $\left(5\dfrac{3}{4} + 1\dfrac{1}{4}\right) - \left(2\dfrac{5}{9} + \dfrac{5}{9}\right)$

32) $50\dfrac{1}{2} - \left(18\dfrac{1}{2} + 5\dfrac{1}{3}\right)$

33) $\dfrac{\dfrac{2}{5}}{\dfrac{3}{8}}$

34) $\dfrac{24}{\dfrac{1}{6}}$

35) $\dfrac{3 + \dfrac{1}{2}}{\dfrac{3}{4}}$

36) $\dfrac{4\dfrac{3}{10} - 2\dfrac{7}{20}}{\dfrac{12}{20} + \dfrac{9}{10}}$

3.5 EXPONENTS AND ORDER OF OPERATIONS WITH FRACTIONS

In simplifying expressions with fractions and mixed numbers, the order of operations still need to be applied. Recall the rules for the order of operations given in previous sections.

> **NOTE:**
> There may be occasions when simplifying with fractions is more straightforward by using the following: $a \div b$ can be rewritten as the fraction $\frac{a}{b}$

PRACTICE PROBLEMS

1) $\left(\dfrac{2}{5}\right)^3$ given expression

     $\dfrac{2}{5} \cdot \dfrac{2}{5} \cdot \dfrac{2}{5}$ simplify exponent

     $= \dfrac{8}{125}$

2) $\dfrac{4}{5} + \dfrac{2}{3} \cdot \dfrac{7}{10}$ given expression

     $\dfrac{4}{5} + \dfrac{7}{15}$ multiply

     $\dfrac{19}{15}$ or $1\dfrac{4}{15}$ add

3) $3\dfrac{1}{10} \div 10 - \left(\dfrac{1}{10}\right)^2$ given expression

     $3\dfrac{1}{10} \div 10 - \dfrac{1}{100}$ simplify exponent

     $\dfrac{31}{100} - \dfrac{1}{100}$ divide

     $\dfrac{30}{100} = \dfrac{3}{10}$ subtract and reduce

4) $\left(\dfrac{1}{4}\right)^3 \div \dfrac{3}{16} + \dfrac{3}{8} \cdot 2\dfrac{1}{4}$ given expression

$\dfrac{1}{64} \div \dfrac{3}{16} + \dfrac{3}{8} \cdot 2\dfrac{1}{4}$ simplify exponent

$\dfrac{1}{12} + \dfrac{3}{8} \cdot 2\dfrac{1}{4}$ divide

$\dfrac{1}{12} + \dfrac{27}{32}$ multiply

$\dfrac{89}{96}$ add

5) $\left(5 \div 15 + \dfrac{2}{3}\right) \cdot \left(4\dfrac{1}{6} - 3\dfrac{5}{6}\right)$ given expression

$\left(\dfrac{5}{15} + \dfrac{2}{3}\right) \cdot \left(4\dfrac{1}{6} - 3\dfrac{5}{6}\right)$ rewrite division as a fraction

$(1) \cdot \left(\dfrac{1}{3}\right)$ simplify within both parenthesis

$\dfrac{1}{3}$ multiply

6) $\left(\dfrac{2}{5} + \dfrac{4}{5}\right)^3 - (12 \div 25)$ given expression

$\left(\dfrac{6}{5}\right)^3 - (12 \div 25)$ add within parenthesis

$\left(\dfrac{216}{125}\right) - \dfrac{12}{25}$ simplify exponent and rewrite as fraction

$\dfrac{156}{125} = 1\dfrac{31}{125}$ subtract

EXERCISE 3.5 SET A

Find each answer. Follow the order of operations.

1) $\left(\dfrac{5}{6}\right)^2$

2) $\dfrac{4}{5} \cdot \left(\dfrac{5}{4}\right)^2$

3) $\left(\dfrac{4}{8}\right)^3 \left(\dfrac{2}{6}\right)^2$

4) $\left(\dfrac{3}{5} \cdot \dfrac{20}{36}\right)^2 \cdot \left(\dfrac{4}{3} \cdot \dfrac{3}{4}\right)$

5) $\left(\dfrac{34}{55}\right)^0 \cdot (4)^3 \cdot \left(\dfrac{1}{8}\right)^2$

6) $\left(3\dfrac{1}{2}\right)^2 + \left(\dfrac{4}{3}\right)^2$

7) $\left(\dfrac{9}{10} + 2\right) \div \left(\dfrac{1}{5} \div 5\right)$

8) $\left(\dfrac{6}{7} - \dfrac{2}{7}\right) \cdot \dfrac{3}{4} + 4\dfrac{2}{3}$

9) $(6 \div 12) + (3 \div 4 \cdot 2)$

10) $\left(\dfrac{3}{8} \div 2\dfrac{3}{4}\right) \cdot \dfrac{1}{4} \, (2 \div 6)$

11) $10\dfrac{3}{8} - \dfrac{1}{4} \div \dfrac{3}{8} \cdot \dfrac{1}{2} + 2\dfrac{1}{8}$

12) $\left(\dfrac{1}{2}\right)^4 \div \left(\dfrac{1}{4}\right)^2 + \left(\dfrac{3}{4} + 3 \div 4\right)$

EXERCISE 3.5 SET B

Find each answer. Follow the order of operations.

1) $\left(\dfrac{3}{8}\right)^2$

2) $\dfrac{6}{7} \cdot \left(\dfrac{7}{6}\right)^2$

3) $\left(\dfrac{5}{10}\right)^3 \left(\dfrac{1}{10}\right)^2$

4) $\left(\dfrac{4}{6} \cdot \dfrac{24}{36}\right)^2 \cdot \left(\dfrac{14}{13} \cdot \dfrac{13}{14}\right)$

5) $\left(\dfrac{1}{9}\right) \cdot (3)^3 \cdot \left(\dfrac{15}{16}\right)^0$

6) $\left(2\dfrac{1}{2}\right)^3 + \left(\dfrac{5}{2}\right)^2$

7) $\left(\dfrac{11}{36} + 2\right) \div \left(\dfrac{1}{6} \div 6\right)$

8) $\left(\dfrac{4}{9} - \dfrac{2}{9}\right) \cdot \dfrac{3}{5} + \left(2\dfrac{2}{3}\right)^2$

9) $(5 \div 15) + (3 \div 4 \cdot 2)$

10) $4\left(\dfrac{1}{8} \div 2\dfrac{3}{4}\right) + (2 \div 6)$

11) $13\dfrac{5}{8} + 1\dfrac{1}{4} \div \dfrac{3}{8} \cdot \dfrac{1}{2} - 2\dfrac{1}{8}$

12) $\left(\dfrac{1}{3}\right)^4 \div \left(\dfrac{1}{9}\right)^2 + \left(\dfrac{2}{3} + 3 \div 4\right)$

3.6 APPLICATIONS WITH FRACTIONS

The numbers in the word problems of this section are fractions and mixed numbers. However, the problem solving model given in a previous section still applies in solving the word problems.

> **POLYA MODEL: 4 STAGES IN PROBLEM SOLVING**
> 1. Read the problem, more than once if needed. As you read the problem, ask yourself what is the given information and is that all the facts.
> 2. Devise a plan or strategy to determine the operations needed to solve the problem.
> 3. Solve the problem. Do the computation from your plan/strategy.
> 4. Check your solution. Did you answer the question? Does the answer make sense in relation to the given information?

PRACTICE PROBLEMS

1) Aleshia has $2\frac{3}{8}$ liters of motor oil. Eugene has $4\frac{3}{8}$ liters of motor oil. Rita has $3\frac{1}{8}$ liters of motor oil. How many liters of motor oil do they have in all?

We will use the Polya Model:

Step 1) The given information tells us how much motor oil each person has. All the amounts are in the same unit — liters.

Step 2) The phrase "in all" indicates we want a total and therefore, the operation is to add the quantities.

Step 3) Do the addition: $2\frac{3}{8} + 4\frac{3}{8} + 3\frac{1}{8} = 9\frac{7}{8}$.

Step 4) The question was answered.

2) Mary had 12 pounds of soil in her garage. She used $\frac{2}{3}$ of it to plant her tulips. How many pounds of soil does she have left?

Step 1) The given information tells us the total amount of soil and the fractional part of the soil used.

Step 2) The question asks how much is left which indicates we need to subtract. However, we were not given the amount of soil that was used. The phrase "$\frac{2}{3}$ of it" is a fractional part of total which indicates we need to multiply to find the amount used. Then subtract to find the amount left.

Step 3) Do the multiplication: $\frac{2}{3} \cdot 12$ pounds = 8 pounds of soil used.

Do the subtraction: 12 pounds − 8 pounds used = 4 pounds left.

Step 4) The question was answered.

Name _____   Date _____

EXERCISE 3.6 SET A

DIRECTIONS:

Write answers with money in appropriate form.
Reduce fractions/ratios to lowest terms.
Round decimals to nearest hundredth/cent.

1) One hundred forty people were surveyed about the genre of movies they like. Ninety people said they liked comedies. What fraction of the people surveyed liked comedies?

2) Lois has 350 coins in her jar. Two hundred of the coins are dimes. What fraction of the coins are dimes?

3) On a 35 multiple choice history test, Lyle answered 25 correctly. What fraction of the test questions did Lyle answer incorrectly?

4) The April issue of We Rock magazine had a total of 1200 square inches of advertisements. Four hundred square inches at $10 per square inch were used to advertise albums. Three hundred square inches at $12 per square inch were used to advertise music stores. Five hundred square inches at $18 per square inch were used to advertise other products such as sodas.

   a) What fractional part of advertisements was used to advertise music stores?

   b) Going To The Dogs used 75 square inches to advertise its new album. What fractional part of advertising albums did Going To The Dogs use?

   c) The Get Your Instrument store used 155 square inches to advertise the Grand Opening of its store in Bloomfield. How much did the store pay for advertising?

5) It takes $\frac{2}{5}$ gram of lead to produce a certain bolt. Express that amount in tenths of a gram.

6) In a cereal packaging plant, machine A fills a box in $3\frac{2}{3}$ seconds, and machine B fills a box in $3\frac{6}{7}$ seconds.

   a) Write the two amounts with a least common denominator.

   b) Which machine fills the box faster?

7) Standards at a fertilizer plant require that each super size bag weigh no less than 100 pounds, but no more than $100\frac{3}{5}$ pounds. An inspector randomly weighs three bags. Bag #1 weighed $100\frac{2}{3}$ pounds. Bag #2 weighed $100\frac{3}{7}$ pounds. Bag #3 weighed $100\frac{1}{2}$ pounds.

   a) Determine which bag of fertilizer weighs more using the least common denominator.

   b) Determine if each bag of fertilizer meets the standards.

8) Terrence bought 5 pounds of cement from a local hardware store. He then bought $5\frac{1}{2}$ pounds more a week later. How many pounds of cement does Terrence now have in all?

9) Roger is the supply manager for a chemistry lab at a local college. At the beginning of each week Roger must have 20 16-liter bottles of saline solution. At the end of last week, Roger inventoried 15 full bottles and 10 bottles that were half full. Did Roger have to order any more saline solution for next week?

10) In order to form 45 grams of a particular solution, Jayson needs 25 grams of carbon. In the supply room, Jayson has three partially filled containers of carbon of the amounts, $12\frac{3}{4}$ grams, $10\frac{1}{6}$ grams, and 14 grams.

   a) How much carbon does Jayson have in the supply room?
   b) Another chemist will lend Jayson $11\frac{1}{3}$ grams of carbon. How many grams does Jayson now have in total?
   c) With the loan, will Jayson have enough carbon to make two solutions?

11) Your kitchen sink's pipe cracked and you need to replace it. You need $5\frac{4}{5}$ inches of copper pipe and $3\frac{1}{4}$ inches of steel pipe. What is the total length of the pipe?

12) Your drainage pipe leading from the garage to the street sewer cracked and needs to be replaced. You will need 12 pipes, each $3\frac{1}{4}$ feet long. You also need a joint to connect the pipes, each $1\frac{1}{2}$ feet long.

   a) How many joints will you need?
   b) What is the total length of the pipe and joints?

13) You expect to place three $2\frac{1}{2}$ feet planks of wood from the edge of the patio to the children's pool. What is the total length?

14) Presently your heating vent is connected by two ducts, $4\frac{3}{4}$ feet and $3\frac{1}{5}$ feet. The house inspector made the recommendation to replace the two ducts with only one duct. What will be the length of the one duct?

15) The apartment over your garage has two types of ducts. The wide ducts, $4\frac{2}{3}$ feet, $3\frac{1}{3}$ feet, and $3\frac{1}{6}$ feet, lead to the kitchen. The narrow ducts, $2\frac{1}{5}$ feet, $5\frac{3}{10}$ feet, and $3\frac{1}{2}$ feet, are found in the kitchen.

   a) What is the total length of wide ducts?
   b) What is the total length of narrow ducts?
   c) The town's fire marshall has indicated that you need to add a $\frac{3}{5}$-feet joint on both sides of each wide duct. What will be the total length of the wide ducts?

16) As their winter hobby, your son and daughter build model cars and airplanes. You will need to add more shelving in their bedrooms to display their models. In Alice's room, you built a new shelf and extended the lengths of the two old shelves to the length of the new shelf. In Mark's room, you built two new shelves. The new shelf is $9\frac{3}{8}$ feet. The old shelves were each $7\frac{1}{4}$ feet.

   a) How many feet of shelving did you add to each old shelf in Alice's room?
   b) How many more feet of shelving are there in Alice's room than Mark's room?

17) You are planning to extend the back of your house on the first floor to include an enclosed patio. The length of the house is $85\frac{1}{2}$ feet. The length of the house and enclosed patio is $98\frac{3}{4}$ feet. What is the length of the indoor patio?

18) Leslie lives $25\frac{1}{2}$ miles from her current job. How many miles does she drive each day round trip?

19) Jenna bought $5\frac{1}{2}$ pounds of cherries at $2 per pound. How much did Jenna pay for the cherries?

20) A container holds $4\frac{1}{6}$ quarts of liquid. How many quarts will five containers hold?

21) Quanda wants her name to appear at the bottom of her business card. Her typed name is a length of $1\frac{1}{6}$ inches. The business card is a length of $3\frac{1}{2}$ inches. How many times will Quanda's name appear at the bottom of her business card?

22) Five college friends went on a road trip for spring break. At the start of the trip the odometer read 96,554 miles. At the end of the trip the odometer read 98,384. They filled the gas tank six times with the amounts of $12\frac{1}{2}$ gallons, $14\frac{2}{3}$ gallons, 13 gallons, $13\frac{1}{4}$ gallons, $12\frac{3}{4}$ gallons, and $8\frac{5}{6}$ gallons. The average price per gallon was $2.
   a) How many miles did the friends travel?
   b) How many miles did the car travel per gallon?
   c) The friends decided to pay the gas bill evenly. How much did each friend pay?

23) Michelle, a figure skater, knows she needs to dedicate many hours at practice to compete at the international level. On all seven days of the week Michelle practices her routine at the ice rink for $3\frac{1}{2}$ hours each day. On three days of the week, Michelle takes ballet lessons for $1\frac{1}{2}$ hours each day. On five days of the week, Michelle does cardio and weight training for $2\frac{1}{2}$ hours each day. How many hours during a typical week does Michelle spend on the various activities to become an elite figure skater?

24) How many minutes is $\frac{2}{3}$ of an hour?

25) Barnaby plays on the basketball court for $\frac{3}{4}$ of an hour each day, 5 days a week. How many hours does Barnaby play each year?

26) Amelia plans to use $\frac{3}{5}$ of the 20-pound bag of soil to re-pot her six plants. How many pounds of soil will Amelia use?

27) Kristoff had 50 gallons of ready-to-mix cement. He used $\frac{2}{3}$ of it to make a patio. How many gallons of cement does Kristoff have left?

28) Carol designated $\frac{1}{5}$ of her weekly paycheck to go directly into her retirement fund. Carol's weekly paycheck is $545. How much money goes to Carol's retirement fund each year?

29) In a recent survey, $\frac{4}{5}$ of the people did not want a parking garage to be built in their town. If five thousand fifty-five people were surveyed, how many people did want a parking garage to be built in their town?

30) The local highway committee has decided to place a "Don't Drink and Drive" sign every $\frac{1}{4}$ of a mile. How many signs need to be ordered if there is a 144 mile stretch of highway in their jurisdiction?

31) Yosef has 48 pounds of trail mix. He will make bags of $\frac{3}{4}$ ounces to give to his hiking customers. How many bags will Yosef have to distribute to his customers? *(16 ounces = 1 pound)*

32) Danielle delivers gallons of spring water to businesses. One day, Danielle re-routed her deliveries and found that she reduced her driving time by $\frac{1}{6}$. If Danielle took $5\frac{2}{3}$ hours to make her deliveries, how long will it take her using her new route?

33) Louis decided to install new windows to help lower the electricity bill. He will save $\frac{1}{5}$ of last years costs. Last year he spent a total of $1740. How much money will Louis save this year?

34) Vera decided to buy a car instead of an SUV. By buying a car, Vera will increase her miles per gallon by $\frac{3}{7}$ compared to a SUV. If a SUV gets 23 miles per gallon, then how many miles per gallon will Vera get on a car?

35) In Sarah's town, the cost of 1 kWh (kilowatt hour) is 8¢. Sarah wants to conserve energy and save money so she has instructed her family to follow the guideline listed on the refrigerator. Last month Sarah's family used 2348 kWh. This month they reduced their usage by $\frac{1}{4}$.
   a) How much money did Sarah save?
   b) How much was this month's bill?

36) Kevin bought a new air conditioner which conserved energy by $\frac{1}{3}$. Kevin's usage last month was 1335 kWh.
   a) How many kWh did Kevin use this month?
   b) The cost of one kWh is 19¢. How much did Kevin save by having the new air conditioner?

37) Oscar bought nine pine wood boards, each measuring $3\frac{3}{4}$ feet. How many linear feet of pine wood boards did he buy?

38) To repair the bathroom wall, Henry bought fifteen one-by-fours, each $6\frac{3}{5}$ foot long. How many linear feet of one-by-fours did he buy?

39) To build the pool deck, you bought 22 two-by-four of pine boards, each $6\frac{2}{3}$ foot long. The cost for pine boards is $32\frac{1}{2}$¢ per linear foot.
   a) How many linear feet of pine boards did you buy?
   b) How much did the pine boards cost?

40) Sasha ordered fifteen $6\frac{3}{5}$-foot one-by-four boards. She will use only $14\frac{1}{2}$ of the boards. How many linear feet will she actually use?

41) Nick ordered twelve oak planks, each $8\frac{1}{8}$ foot long. The cost of the oak planks is $35\frac{1}{4}$¢ per linear foot.
   a) Nick used only $11\frac{3}{4}$ of the oak planks. How many linear feet did he actually use?
   b) How much was the cost of the oak planks?

42) Calvin bought twenty-five 6-foot mahogany boards to build a magazine rack. The cost of the mahogany boards is $75\frac{1}{2}$¢ per linear foot. He needed to cut 12 boards to a length of $5\frac{1}{4}$ feet which he will use.
   a) How many linear feet of the boards did he waste?
   b) How much was the cost of the mahogany boards?

43) Wanda bought thirty plaster wallboards, each a length of 5 foot. The cost of the plaster wallboards is $45\frac{2}{3}$¢ per linear foot. She will need to coat the wallboards with flat white paint. Each wallboard requires $1\frac{1}{4}$ gallons of paint. The cost of a gallon of paint is $6.
   a) What is the total cost for the plaster wallboards and the paint?
   b) The Lets Fix It Hardware store sells the plaster wallboards already painted at $14 per wallboard. How much will Wanda save by doing it herself?

44) As a telemarketer, Nora gets paid $10 an hour for a 35-hour week and $1\frac{1}{2}$ times that rate for any hours worked over 35 hours. For the past 3 weeks, Nora has been asked to work overtime. For the first week, Nora worked a total of 40 hours. For the second week, Nora worked a total of 43 hours. For the third week, Nora worked a total of 45 hours. What was Nora's total pay at the end of the 3 weeks?

45) Rachel needs a total of 5 pounds of various cheeses to make her famous four cheese baked ziti. Rachel bought $2\frac{1}{2}$ pounds of mozzarella, $1\frac{1}{4}$ pounds of white cheddar, and $\frac{3}{4}$ pounds of Swiss. How many pounds of Gouda cheese does Rachel need to buy to make her ziti dish?

46) Gordon wants to enclose his rectangular garden with a white picket fence. The length of the garden is 5 feet and the width is $7\frac{3}{5}$ feet.
   a) How many feet of fencing does Gordon need to buy?
   b) At Put Up Fences store, white picket fences sell for $65 per foot. How much did Gordon pay for his fence?

47) Walter has a 12-foot board to make a mailbox. He needs to cut the board into $\frac{3}{4}$ foot long pieces. How many pieces of boards can Walter cut?

48) Bart, the tight end on his high school football team, weighs $280\frac{1}{4}$ pounds. Alberta, a gymnast on her high school team, weighs $112\frac{1}{10}$ pounds. How many times more does Bart weigh than Alberta?

49) Nahomi received $3\frac{1}{2}$ dozen roses while she was in the hospital recuperating from knee surgery. She kept $\frac{2}{3}$ of them in her room and gave the rest of them to another patient down the hall. How many roses did Nahomi give to the patient?

Name _____   Date _____

EXERCISE 3.6 SET B

DIRECTIONS:

Write answers with money in appropriate form.
Reduce fractions/ratios to lowest terms.
Round decimals to nearest hundredth/cent.

1) The Excal Company has 48 employees. Thirty of the employees are female. What fraction of the employees are females?

2) A local company with 200 employees wants to start up a softball team or a bowling team. In a survey, 145 employees wanted to start a bowling team. What fraction of the employees wanted to start a softball team?

3) The Kleats Arts Theatre has 75 orchestra seats at $50 per seat, 12 balcony seats at $47 per seat, 100 lower level seats at $30 per seat, and 200 upper level seats at $32 per seat.
   a) What fraction of all the seats are orchestra seats?
   b) The orchestra and balcony seats are considered the best seats. What fraction of the best seats are balcony seats?
   c) The production of "The Wizard of Oz" was sold out for Friday night, Saturday night and Sunday afternoon and evening. What was the total amount made from ticket sales?

4) A cupcake company requires $\frac{3}{4}$ ounces of sugar for each cupcake. Express that amount in sixteenths of an ounce.

5) A fabric softener company added $3\frac{7}{10}$ milliliters of liquid to each liter of fabric softener. Express that amount in hundredths of a milliliter.

6) The thickness of a piece of oak wood is $\frac{12}{5}$ inch and the thickness of a piece of pine wood is $2\frac{3}{4}$ inch.
   a) Write the two amounts with a least common denominator.
   b) Which wood is thinner?

7) A dish detergent company requires that each bottle of dish detergent be filled with no less than 32 ounces, but no more than $32\frac{1}{3}$ ounces. Every hour an inspector randomly chooses two bottles to check if they meet the requirements. In one instance, bottle #1 weighed $32\frac{1}{4}$ ounces and bottle #2 weighed $32\frac{3}{8}$ ounces.
   a) Write both weights as an improper fraction.
   b) Write the two improper fractions with a least common denominator. Determine which bottle contains more dish detergent.
   c) Determine if each bottle meets the requirements set by the company.

8) Harry has $3\frac{5}{7}$ grams of sodium. Ron has $4\frac{3}{7}$ grams of sodium. Ginny has $5\frac{6}{7}$ grams of sodium. How many grams of sodium do they have in all?

9) You are the head chef of desserts at a local restaurant. You need 100 pounds of sugar to make the desserts tonight. There are five un-opened bags of $15\frac{1}{2}$ pounds each. There are three opened bags of 12 pounds, $4\frac{3}{4}$ pounds, and $8\frac{1}{2}$ pounds. Do you have enough sugar to make all the planned desserts?

10) Keith has $6\frac{3}{5}$ gallons of shampoo in one bottle, and $1\frac{1}{4}$ gallons in another bottle in his supply cabinet. In her supply cabinet, Samira has $5\frac{3}{4}$ gallons of shampoo in one bottle, and $2\frac{1}{8}$ gallons in another bottle

   a) If Keith combines both bottles into one, how many gallons of shampoo will he have?
   b) If Samira combines both bottles into one, how many gallons of shampoo will she have?
   c) The inventory manager for the hair salon recommends that there always be a total of 15 gallons. If Keith and Samira combine their amounts, will they meet the recommendation?

11) Bobby bought 3 copper pipes, each $7\frac{1}{6}$ inches, and 4 plastic pipes, each $\frac{8}{9}$ inches.
    a) What is the total length of the copper pipes?
    b) What is the total length of plastic pipes?

12) Peter has a 30-inch length of pipe. He needs to cut it into 3 equal lengths of $9\frac{9}{10}$ inches. Is the pipe long enough?

13) The manager of Unit 26 found there are 6 pipes leading to each apartment in the unit. Pipe #1 and pipe #2 are $5\frac{2}{3}$ feet and $3\frac{1}{4}$ feet, respectively. Pipe #3 is two feet more than pipe #1 and pipe #2 combined. Pipe #4 and pipe #5 have the same length. Pipe #4 is $1\frac{1}{5}$ feet. Pipe #6 is as long as pipe #2, #3, and #4.
    a) What is the total length of pipe #1 and pipe #2 combined?
    b) What is the total length of pipe #3 and pipe #5?
    c) What is the total length of pipe #6?

14) You want to add a walk-in closet to your bedroom. The length of your bedroom is $15\frac{1}{2}$ feet and the width is $16\frac{3}{4}$ feet. You got an estimate from a construction manager. He says you can add the walk-in closet by extending the bedroom's length to $16\frac{3}{4}$ feet and the bedroom's width to $17\frac{1}{4}$ feet. He will charge $250 for materials and $45 per hour of labor for each construction worker.
    a) How much longer will the bedroom be?
    b) How much wider will the bedroom be?
    c) If there are two construction workers assigned to this job and they worked for a total of 20 hours throughout 3 days, how much will it cost to add the walk-in closet?

15) Since your love for growing vegetables has grown over the years, you decided to extend the length of your garden from $4\frac{1}{2}$ feet to $10\frac{2}{5}$ feet.

    [--- $3\frac{2}{3}$ ft -----][----- $4\frac{1}{2}$ ft ----------][--------- ?---------]

	original garden	

    a) How much longer is the new garden than the original one?
    b) The garden was not extended evenly on each end. The garden was extended $3\frac{2}{3}$ feet on the left side. What length was the garden extended on the right side?

16) Your car gets $34\frac{1}{4}$ miles per gallon. You filled up your tank twice this week with the amounts of 12 gallons each time. How many miles did you drive?

17) Three hundred fifty people accepted the invitation to Powell and Barbara's wedding. Only 260 people attended because of a blizzard. What fractional part of the guests did not attend the wedding?

18) A recipe for vegetable lasagna yields 6 servings and needs $3\frac{1}{2}$ cups of ricotta cheese. Tania will make the lasagna to yield 12 servings. How many cups of ricotta cheese will she need?

19) Geraldo has $5\frac{3}{4}$ pounds of sunflower seeds. He will distribute the sunflower seeds to 10 people. How many pounds will each person receive?

20) On average, Wesley's routine has been the same since his retirement. On Monday, Wednesday, and Friday, he drives $2\frac{1}{2}$ miles to the local library, then $8\frac{1}{4}$ miles to the diner to meet his other friends in retirement. How many miles does Wesley drive roundtrip each week?

21) Jim has to paint each of his children's room with their color of choice. He purchased three cans of lime green paint, each $4\frac{1}{2}$ gallons, for his daughter. He purchased three cans of hot-rod blue paint, each $3\frac{3}{4}$ gallons, for his son. Each gallon of paint costs $7 per gallon. How much in total did Jim spend on paint?

22) "We Serve You" Caterers have been hired to cater a small party of 25 people. One of the dinner items is stuffed shells. The head chef allows $\frac{3}{4}$ ounces of ground beef for each stuffed shell and plans for each person to have 3 stuffed shells. How many ounces of ground beef does the head chef need to order?

23) Let's Farm Inc. has $69\frac{1}{2}$ acres of land. Four and one-fourth acres have been set aside for roads and buildings. With the remaining land, Let's Farm Inc. will subdivide it into $7\frac{1}{4}$-acre plots. How many plots will there be?

24) Terrence drove $\frac{3}{8}$ of his 200-mile trip on Monday. How many miles did Terrence drive on Monday?

25) Janice saved $1449 in the past year. She used $\frac{5}{7}$ of her savings to buy a plane ticket to England. How much money does she have left in her savings?

26) Orley took out a student loan for each year he was in college. The amounts were $985, $1055, $1120, $2885, and $780. Five years after graduation, Orley paid off $\frac{5}{8}$ of his total loan amount. How much is left for Orley to pay on his loans?

27) The force of gravity on the Moon is about $\frac{1}{6}$ that of the Earth. How much would a 160 pound astronaut weigh on the moon?

28) At the end of the year Rowena recorded that she walked a total of 1430 miles. If Rowena walked five days a week, how many miles did she walk each day?

29) There are 16 ounces in one pound. How many ounces are in $12\frac{3}{4}$ pounds?

30) By taking a shorter route to work, you found that you can reduce the number of miles by $\frac{1}{5}$. You averaged 45 miles per day. How many miles less do you drive now that you take the shorter route?

31) You are determined to save some money this year by being conservative on water and electricity usage. You estimate that you will save $\frac{1}{3}$ of the water bills and $\frac{2}{5}$ of the electricity bills. If last year you spent $1224 on your water bill and $1860 on your electricity bill, how much combined will you save this year?

32) Romano has decided to install insulation in his attic. This will decrease his heating bill by $\frac{1}{4}$. Last year Romano's heating bill during the winter months of December, January, and February averaged $164 per month. How much will Roman pay each month this year now that he has the insulation in the attic?

33) When you purchased your new home, you kept the dishwasher which used 2400 kWh per month. After one year you were able to buy a new dishwasher which reduced the kWh usage by $\frac{3}{8}$. The cost of electricity is 12¢ per kWh.
   a) How much kWh do you use with your new dishwasher?
   b) How much money will you save each year with the new dishwasher?

34) To build a tree house for your children, you will need 12 one-by-four oak wood planks, each $5\frac{1}{2}$ foot long. How many linear feet of oak wood planks will you need?

35) Leann bought eleven $6\frac{4}{5}$-foot lengths of two-by-four oak boards to build a birdhouse. The cost of the oak boards is $25\frac{3}{4}$¢ per linear foot.
   a) How many linear feet of oak boards did Leann buy?
   b) How much did Leann pay for the lumber?

36) Tony needs 24 oak boards, each $9\frac{2}{3}$ foot long, to build a doghouse. He already has 16 oak boards of the right length in the garage.
   a) How many linear feet does Tony need?
   b) How many linear feet does Tony have in the garage?
   c) How many more linear feet does Tony need to buy?

37) Johnny ordered eight $4\frac{1}{4}$-foot lengths of pine boards to build shelves. He will use only $7\frac{1}{3}$ of the pine boards. How many linear feet will he actually use?

38) Georgia ordered fourteen 5-foot lengths of two-by-four boards. She will actually use only $13\frac{1}{4}$ of the boards. The cost of the two-by-four boards is $50\frac{1}{2}$¢.
   a) How many linear feet will Georgia use?
   b) How much was the cost of the boards?

39) Zachary bought ten $9\frac{2}{3}$-foot oak boards at Hometown Hardware store at $65\frac{1}{2}$¢ per linear foot. In the Sunday paper, Zachary saw the All-Around Needs Hardware store advertises oak boards at $61\frac{3}{4}$¢ per linear foot. How much would Zachary have saved if he had bought the oak boards at All-Around Needs Hardware?

40) Nelly bought twenty 8-foot lengths of oak boards to build a bookcase. The cost of the oak boards is $68\frac{3}{4}$¢ per linear foot. He needed to cut fifteen of the oak boards to a length of $7\frac{1}{3}$ feet and the remaining five of the oak boards to a length of $7\frac{1}{4}$ feet which he will use.
   a) How many linear feet of the boards did he waste?
   b) How much was the cost of the oak boards?

41) Ben considers himself a handyman and makes things from scratch. His twin brother, Ken, buys everything ready made. Both Ben and Ken want a deck for their home. Ben buys forty 6-foot oak boards at $67\frac{1}{2}$¢ and the 10 gallons of stain finish at $10 per gallon. Ken hires a local carpenter who charges $120 for materials and $220 for labor.
   a) How much did it cost Ben to build the deck?
   b) How much more did Ken pay to have someone build his deck than what it cost his brother, Ben?

42) African elephants can weigh up to 320 pounds when they are born. When they reach adulthood, the African elephants can weigh up to $37\frac{1}{2}$ times its weight at birth. How many pounds will an African elephant weigh by adulthood?

43) Over four days the Mansley's drove 1200 miles across country. On the first day, they drove $\frac{1}{5}$ of the miles. On the second day, they drove $\frac{1}{3}$ of the remaining miles. On the third day, they drove $\frac{2}{5}$ of the remaining miles. How many more miles did the Mansley's have left to drive on the fourth day?

44) Margo earns $56,345 a year. One-twentieth of her annual salary goes toward savings. How much money does Margo save each year?

45) Alexi's famous champagne punch fountain requires $1\frac{3}{4}$ quart of orange sherbert. This weekend he will be catering 3 weddings. At each of the 3 weddings, Alexi will display 5 champagne punch fountains throughout the catering hall. How many quarts of orange sherbert will Alexi need to order?

46) Massimo had $3\frac{1}{2}$ pounds of jelly beans. He kept $\frac{2}{3}$ pounds to put in his candy dish at work and gave the rest of them to a co-worker. How many pounds did Massimo give his co-worker?

47) Zoran's goal is to walk $3\frac{1}{2}$ miles five times a week and jog $1\frac{1}{2}$ miles two times a week. If Zoran keeps his goal, how many miles will he have walked and jogged in a year?

48) Vidal has a rectangular oak board with a length of $2\frac{3}{4}$ feet and a width of $3\frac{1}{2}$ feet. Vidal will cover the oak board with striped silk fabric costing $25 per square foot. What is the total cost of the fabric?

49) Julliard School in New York City requires an audition before a student is accepted. Evan would like to attend Julliard School. For one year before his audition, Evan increased practicing the piano from one hour a day, three times a week to $2\frac{1}{2}$ hours, 6 days a week. How many more hours did Evan practice that year?

50) Yum-Yum Bakery sells Pinoli nuts for $7 per pound. Orphelia will make fifteen of her famous banana nut cakes with Pinoli nuts for the bake sale. Each cake requires $\frac{1}{4}$ pound of nuts.
    a) How many pounds of Pinoli nuts does Orphelia need to buy?
    b) How much was the cost of the Pinoli nuts?

51) The Ecology Club of a local high school decided to place a sign announcing their car wash event. A sign was placed every $\frac{1}{5}$ mile on a 5-mile stretch of highway. How many signs does the Ecology Club need to make?

3.7 EXPONENTS AND ORDER OF OPERATIONS WITH SIGNED FRACTIONS

Within this chapter, we have simplified expressions with fractions and mixed numbers. Each of the fraction or mixed number was a positive number. We need to keep in mind that we can simplify positive or negative fractions and mixed numbers. The order of operations and the rules for operations with signed numbers will still hold true.

Recall the rules for operations with signed numbers.

> **Addition or Subtraction Rules for Signed Numbers:**
> - If the numbers have the same signs, add the numbers and keep the common sign.
> - If the numbers have different signs, subtract the numbers and keep the sign of the larger number.
>
> **Multiplication or Division Rules for Signed Numbers:**
> 1) Perform the indicated operation, multiplication or division.
> 2) Determine the sign of the answer.
>   - If the numbers have the same sign, the answer will be a positive number.
> $(+ \cdot + = +; - \cdot - = +; + \div + = +; - \div - = +)$
>   - If the numbers have different signs, the answer will be a negative number.
> $(+ \cdot - = -; + \div - = -)$

NOTE:

A negative fraction can be written in three equivalent ways:

$$-\frac{a}{b} = \frac{-a}{b} = \frac{a}{-b}$$

PRACTICE PROBLEMS

1) $\dfrac{14}{15} - \dfrac{23}{15} = \dfrac{14}{15} + \left(-\dfrac{23}{15}\right) = -\dfrac{9}{15} = -\dfrac{3}{5}$ (different signs: subtract and keep sign of larger number)

2) $-\dfrac{5}{11} - \dfrac{6}{11} - \dfrac{9}{11} = -\dfrac{20}{11}$ or $-1\dfrac{9}{11}$ (same signs: add and keep the common sign)

3) $3\dfrac{1}{2} - \left(-\dfrac{4}{5}\right) = 3\dfrac{1}{2} + \dfrac{4}{5} = 4\dfrac{3}{10}$ or $\dfrac{43}{10}$ (recall: $-(-a) = a$)

4) $-\dfrac{18}{25} \div -\dfrac{9}{5} = -\dfrac{18}{25} \cdot -\dfrac{5}{9} = \dfrac{2}{5}$ (divide with same signs, answer is positive)

5) $\left(-\dfrac{2}{5}\right)^3 = -\dfrac{2}{5} \cdot -\dfrac{2}{5} \cdot -\dfrac{2}{5} = -\dfrac{8}{125}$ (multiply with signs; $- \cdot - = +$, then $+ \cdot - = -$)

6) $\dfrac{4}{5} + \dfrac{2}{3} \cdot -\dfrac{7}{10}$ given expression

$\dfrac{4}{5} + \left(-\dfrac{7}{15}\right)$ multiply $(+ \cdot - = -)$

$\dfrac{1}{3}$ different signs: subtract and keep sign of larger number

7) $-3\dfrac{1}{10} \div 10 - \left(\dfrac{1}{10}\right)^2$ given expression

$-3\dfrac{1}{10} \div 10 - \dfrac{1}{100}$ simplify exponent

$-\dfrac{31}{100} - \dfrac{1}{100}$ divide; $(- \div + = -)$

$-\dfrac{32}{100} = -\dfrac{8}{25}$ same signs: add and keep common sign

8) $-\left(-\dfrac{1}{4}\right)^3 \div -\dfrac{3}{16} + \dfrac{3}{8} \cdot -2\dfrac{1}{4}$ given expression

$-\left(-\dfrac{1}{64}\right) \div -\dfrac{3}{16} + \dfrac{3}{8} \cdot -2\dfrac{1}{4}$ simplify exponent (negative number to an odd power is negative)

$\dfrac{1}{64} \div -\dfrac{3}{16} + \dfrac{3}{8} \cdot -2\dfrac{1}{4}$ simplify $-(-a) = a$

$-\dfrac{1}{12} + \dfrac{3}{8} \cdot -2\dfrac{1}{4}$ divide (different signs, answer is negative)

$-\dfrac{1}{12} - \dfrac{27}{32}$ multiply (different signs, answer is negative)

$-\dfrac{89}{96}$ same signs; add and keep common sign

Name _____ Date _____

EXERCISE 3.7 SET A

Find each answer. Follow the order of operations.

1) $\left(-\dfrac{5}{6}\right)^3$

2) $-\dfrac{4}{5} \cdot \left(\dfrac{5}{4}\right)^2$

3) $\left(-\dfrac{4}{8}\right)^3 \left(\dfrac{2}{6}\right)^2$

4) $\left(-\dfrac{3}{5} \cdot \dfrac{20}{36}\right)^2 \cdot \left(-\dfrac{4}{3} \cdot \dfrac{3}{4}\right)$

5) $\left(-\dfrac{34}{55}\right)^0 \cdot (4)^3 \cdot -\left(\dfrac{1}{8}\right)^2$

6) $\left(3\dfrac{1}{2}\right)^2 - \left(\dfrac{3}{4}\right)^2$

7) $\left(\dfrac{9}{10} - 2\right) \div \left(\dfrac{1}{5} \div -5\right)$

8) $\left(-\dfrac{4}{7} - \dfrac{2}{7}\right) \cdot \dfrac{3}{5} + (2)^2$

9) $(-6 \div 12) + (-3 \div 4 \cdot -2)$

10) $-\left(\dfrac{3}{8} \div \dfrac{3}{4}\right) - \dfrac{1}{4}(2 \div -6)$

11) $-24\dfrac{3}{8} - \dfrac{1}{4} \div \dfrac{3}{8} \cdot -\dfrac{1}{2} + 2\dfrac{1}{8}$

12) $\left(-\dfrac{1}{2}\right)^3 \div \left(\dfrac{1}{4}\right)^2 + \left(\dfrac{3}{4} - 3 \div 4\right)$

EXERCISE 3.7 SET B

Find each answer. Follow the order of operations.

1) $-\left(-\dfrac{3}{8}\right)^2$

2) $-\dfrac{6}{7} \cdot \left(\dfrac{7}{6}\right)^2$

3) $\left(\dfrac{-5}{10}\right)^3 \left(\dfrac{1}{10}\right)^2$

4) $\left(\dfrac{4}{6} \cdot \dfrac{24}{36}\right)^2 \cdot \left(\dfrac{14}{13} \cdot -\dfrac{13}{14}\right)$

5) $\left(\dfrac{1}{9}\right) \cdot -(3)^3 \cdot \left(\dfrac{15}{16}\right)^0$

6) $\left(-2\dfrac{1}{2}\right)^3 - \left(\dfrac{5}{2}\right)^2$

7) $\left(-\dfrac{11}{36} + 2\right) \div \left(-\dfrac{1}{6} \div 6\right)$

8) $\left(\dfrac{4}{9} - \dfrac{2}{9}\right) \cdot \dfrac{3}{5} - \left(\dfrac{2}{3}\right)^2$

9) $(-5 \div 15) + (-3 \div 4 \cdot 2)$

10) $-4\left(\dfrac{1}{8} \div 2\dfrac{3}{4}\right) - (-2 \div 6)$

11) $-13\dfrac{5}{8} - 1\dfrac{1}{4} \div \dfrac{3}{8} \cdot \dfrac{1}{2} - 2\dfrac{1}{8}$

12) $-\left(\dfrac{1}{3}\right)^4 \div \left(\dfrac{1}{9}\right)^2 + \left(-\dfrac{2}{3} + 3 \div 4\right)$

3.8 MORE WORDS TO ALGEBRA

In an earlier section, you were given a list of key words for operations and the algebraic translation. We will use the same key words though the numbers may be fractions and mixed numbers. Below is an abbreviated version to refresh your memory.

Computation	Key Word(s)	Computation	Key Word(s)
Addition	plus add sum added to more than increased by total	Multiplication	times multiplied by product twice *(times 2)* doubled *(times 2)* tripled *(times 3)* fractional part of
Subtraction	minus subtract from difference fewer than less than decrease by	Divide	divided by quotient
Equality	is equal to is result is same result as		

PRACTICE PROBLEMS

Write each sentence as an algebraic expression. The variable "x" will be used for the unknown.

1) the sum of twelve and a number $\qquad$ $12 + x$

2) the product of two-thirds and a number $\qquad$ $\dfrac{2}{3} \cdot x$

3) four-fifths of the difference of a number and seven $\qquad$ $\dfrac{4}{5}(x - 7)$

4) the quotient of 5 and a number is one-half of the number $\qquad$ $\dfrac{5}{x} = \dfrac{1}{2}x$

Name _____   Date _____

EXERCISE 3.8 SET A

Write each sentence as an algebraic expression or equation.

1) The sum of six and a number

2) The difference of eight and one-half a number

3) The product of a number and three-fourths

4) The quotient of a number and fifteen

5) Five less than three-fifths of a number

6) The product of three-fourths of a number and eleven

7) The quotient of three less than a number and four

8) One more than eight-ninths of a number

9) Twice the sum of one-third and a number

10) Two-sevenths of a number, decreased by one-seventh of the number

11) Thirteen plus one-half of a number is three less than the number.

12) The sum of twenty and twice a number gives the same result as one-half the number.

13) The difference of ten times a number, six times the number, and four times the number is five.

14) The quotient of a number and −4, increased by twelve, is twenty-one.

15) Three-fourths the sum of a number and its reciprocal is thirty-three.

16) The product of one-fifth a number, two-fifths the number, and three-fifths the number is fifty.

17) Forty-five less than three times a number is equal to one-half the number.

18) Two-ninths of a number, divided by two is the same result as five more than the number.

19) Six times the sum of fourteen and two-eighths of a number is one.

20) The product of five more than a number and one-fourth is twenty-four.

EXERCISE 3.8 SET B

Write each sentence as an algebraic expression or equation.

1) The sum of a number and -5

2) The difference of two-thirds a number and three

3) The product of a number and one-half

4) The quotient of twelve and a number

5) Six increased by four-fifths of a number

6) The product of three-fourths of a number, one-fourth of the number, and sixteen

7) The quotient of a number decreased by five and eight

8) One less than two-ninths of a number

9) Three times the sum of a number and two-thirds

10) Two-sixths of a number minus one-seventh of the number

11) Thirteen minus one-fourth of a number is three more than one-half the number.

12) The quotient of three times a number and nine gives the same result as one-half the number.

13) The sum of four times a number, seven times the number, and three times the number is three.

14) The quotient of six and a number, decreased by ten, is twenty-two.

15) One-third the sum of a number and its reciprocal is twelve.

16) The product of three-eighths a number, two-eighths the number, and five-eighths the number is ninety.

17) Seventy-two subtracted from twice a number is equal to one-fifth the number.

18) Two-ninths of a number multiplied by two-thirds the number is the same result as five more than the number.

19) Five times the difference of four and two-eighths of a number is fourteen.

20) The product of five less than a number and three more than one-fourth the number is twenty-nine.

4.1 READING AND WRITING DECIMALS

Decimal numbers: Numbers that are gloomy and miserable are dismal numbers

I am sure you have seen numbers such as 15.32, .5, 0.638, 127.009. These kinds of numbers are called **decimal numbers** and the dot used in them is called the **decimal point.** Decimal numbers give another way to express proper fractions and mixed numbers. Let me explain.

In our number system, we have learned place values for whole numbers. The place values get smaller as we go from left to right. Each place value is equal to the previous value divided by 10. So any digit to the right of the ones place must have a value that is less than one—a fraction.

```
                                    hundred-thousands
                  hundred-thousands                  ten-thousandths
         ten-thousands                     hundredths hundred-thousandths
     millions   thousands  hundreds tens ones  thousandths  millionths
        9  ,  4  5  6  ,  1  2  7  .  1  2  7  1· 2  7
                                  and
```

For example, in 0.576,

the 5 means 5 tenths $\left(\dfrac{5}{10}\right)$,

the 7 means 7 hundredths $\left(\dfrac{7}{100}\right)$,

the 6 means 6 thousandths $\left(\dfrac{6}{1000}\right)$.

Numbers to the right of the decimal point are read like whole numbers and are given a value according to the position of its last digit.

```
tenths
 hundredths
  thousandths
   ten-thousandths
```

0.7   is read "seven tenths." $\left(\dfrac{7}{10}\right)$

0.2 3   is read "twenty-three hundredths." $\left(\dfrac{23}{100}\right)$

0.1 7 7   is read "one hundred seventy-seven thousandths." $\left(\dfrac{177}{1000}\right)$

0.0 0 4 8   is read "forty-eight ten-thousandths." $\left(\dfrac{48}{10,000}\right)$

Notice that place values to the right of the decimal point end in "ths."

1.) Pages 009–018, 037–042, 045–047, 052–059, 062–077, 080–087, 115–127, 129–132, 141–143, 147–152, 184–187, 199–203, 206–209, from *Math for College Students: Arithmetic with Introductions to Algebra and Geometry*, 5th Edition by Ronal Straszkow. © 1997 by Kendall/Hunt Publishing Company. Used with permission.

If we have numbers to the right and left of the decimal point, we really have a mixed number. For example, 5.3 is read "five and three tenths."

- five = whole number
- . = decimal point
- three tenths = fractional part

16.08 is read "sixteen and eight hundredths."

172.056 is read "one hundred seventy-two and fifty-six thousandths."

We can summarize **reading numbers** that contain a decimal point as follows:

> 1. Read the whole number part before the decimal point.
> 2. Read an "and" for the decimal point.
> 3. Read the number to the right of the decimal point along with the place value determined by the last digit of the number.

NOTE: Decimal numbers that have no whole number part, such as .6, .37, or .0063, are often written with a zero in the ones place—0.6, 0.37, or 0.0063. The zero in the ones place does two things: it calls attention to the decimal point and it emphasizes that the number is less than one; it has zero ones.

Example 1

Represent 20,056.0073 in words.

twenty thousand, fifty-six and seventy-three ten-thousandths

Example 2

Write one thousand, five and seven hundredths using numbers.

1,005.07

Example 3

In the number 23.75 what does the:

7 represent?	7 tenths $\left(\dfrac{7}{10}\right)$
5 represent?	5 hundredths $\left(\dfrac{5}{100}\right)$
3 represent?	3 ones (3)
2 represent?	2 tens (20)

PROBLEMS

1) Write 215.0705 in words.

Answer: two hundred fifteen and seven hundred five ten-thousandths

2) Write using numbers:
two thousand fifty-five and twelve hundredths.

Answer: 2,055.12

EXERCISE 4.1

In problems 1–7, write each number in words.

_____ 1) 0.5

_____ 2) 0.17

_____ 3) 0.039

_____ 4) 5.0007

_____ 5) 16.35

_____ 6) 426.9

_____ 7) 6.1236

In problems 8–15, write each expression using numbers.

_____ 8) seven tenths

_____ 9) twelve hundredths

_____ 10) nine and three thousandths

_____ 11) forty-five and six ten-thousandths

_____ 12) one hundred and sixteen hundredths

_____ 13) three hundred fifty-six and two hundred seven thousandths

_____ 14) five thousand, twenty-three and three thousand, five hundred seventeen ten-thousandths

_____ 15) two million, eighty thousand and one thousand eighty-six millionths

In the number 167.2354,

_____ 16) What does the 2 represent?

_____ 17) What does the 5 represent?

_____ 18) What does the 4 represent?

_____ 19) What does the 6 represent?

_____ 20) What does the 7 represent?

_____ 21) What does the 3 represent?

4.2 ROUNDING OFF DECIMALS

The more digits there are after the decimal point in a number, the more difficult it is to read and work with that number. In many situations, an approximate or rounded off number would suffice. Rounded off decimal numbers have a value that is close to the original number.

Decimal numbers are rounded off using a similar procedure to the one used to round off whole numbers.

> **ROUNDING OFF DECIMAL NUMBERS**
> 1. Look at the digit to the right of the place that you are rounding off to.
> 2. If that digit is 5 or more, round off by increasing the digit in the place you are rounding to by one, and discarding all the digits to the right of that place.
> 3. If that digit is less than 5 round off by discarding all the digits to the right of the place you are rounding to.

This is made clear by examining some examples.

Example 1

Round off 3.47 to the nearest tenth (to one **decimal place**).

$$3.47 \approx 3.5$$

tenths place ⟶ digit to the right is more than 5

Round off by increasing the tenths place by 1 (changing the 4 to a 5), and discarding the digits to its right.

Example 2

Round off 17.4236 to two decimal places.

$$17.4236 \approx 17.42$$

second decimal place ⟶ digit to the right is less than 5

Round off by discarding the digits to the right of the second decimal place.

Example 3

Round off 0.5895 to the nearest thousandth.

$$0.5895 \approx 0.590$$

thousandths place ⟶ digit to the right is 5

Round off by increasing the thousandths place by one and discarding the digits to the right of the thousandths place.

$0.5895 \approx 0.590$ (*Note:* The zero is left at the end of the answer. Since the problem asks you to round off to the thousandths place, there should be a digit in the thousandths place.)

PROBLEMS

Round off to the nearest:

	whole number	tenth	hundredth	thousandth
1) 4.7658	_____	_____	_____	_____
2) 15.0313	_____	_____	_____	_____
3) 105.47532	_____	_____	_____	_____
4) 68.599512	_____	_____	_____	_____
5) 0.60051	_____	_____	_____	_____

Answers: 1) 5, 4.8, 4.77 4.766
2) 15, 15.0 15.03 15.031
3) 105, 105.5, 105.48 105.475
4) 69, 68.6, 68.60, 68.600
5) 1, 0.6, 0.60, 0.601

Round off 5672.418 to the following:

6) nearest ten

7) nearest hundred

8) nearest thousand.

Answers: 6) 5670

7) 5700

8) 6000

Name _____ Date _____

EXERCISE 4.2

In problems 1–6, round off each number to the nearest tenth.

_____ 1) 36.72

_____ 2) 4.65

_____ 3) 125.482

_____ 4) 89.971

_____ 5) 8.95

_____ 6) 0.149

In problems 7–12, round off each number to the nearest hundredth.

_____ 7) 18.724

_____ 8) 5.475

_____ 9) 792.038218

_____ 10) 0.8964

_____ 11) 7.195

_____ 12) 1.0249

In problems 13–16, round off 562.01846 to the indicated place.

_____ 13) thousandth

_____ 14) ten-thousandth

_____ 15) ten

_____ 16) hundred

In problems 17–19, round off 46.96352 to the indicated place.

_____ 17) 2 decimal places

_____ 18) 3 decimal places

_____ 19) 4 decimal places

4.3 ADD, SUBTRACT, MULTIPLY, DIVIDE WITH DECIMALS

The method for adding decimal numbers is very similar to adding whole numbers. You want to make sure that you add the digits with the same place value.

> **ADDING DECIMALS**
> 1. Place the numbers in a column so that their decimal points are lined up.
> 2. Add the digits that have the same place value.
> 3. Place the decimal point in the answer below the other decimal points.

For example, to add 5.3 + .076 + 12.21:

Line up the decimal points.

```
   5.3
    .076        Add the digits in each column and place the decimal point in the answer below
 +12.21         the other decimal points.
  17.586
```

If any of the numbers you are trying to add are whole numbers without a decimal point, you can simply place a decimal point at the end of the whole number.

For example: 26 = 26.
             5 = 5.
         37,612 = 37,612.

Let's examine a few examples.

Example 1

52 + 3.01 + 0.035 + 1.58 = ?

Line up the decimal points.

```
   52.
    3.01         (Notice a decimal point was placed at the end of 52 before adding.)
    0.035
 +  1.58         Add each column and place the decimal point in the answer below the other
   56.625        decimal points.
```

Example 2

307.52 + 136 + .65 + 28 + 1.17 + 0.2 = ?

```
  307.52
  136.
     .65
   28.
    1.17
 +  0.2
  473.54
```

Subtracting or finding the difference between decimal numbers is similar to adding decimal numbers in that you want to operate on digits with the same place value. To do this you must line up the decimal points before you do the subtraction.

For example: $27.973 - 2.241 = ?$

1. Line up the decimal points.

$$\begin{array}{r} 27.973 \\ -\ 2.241 \\ \hline 25.732 \end{array}$$

Difference:
My washing machine has three different cycles. I call each a differ-ence.

2. Subtract and place the decimal point in the answer below the other points.

In that example, both numbers had 3 decimal digits. Subtraction is easier when both numbers have the same number of decimal digits. In order to have that happen with other subtraction problems, we sometimes have to give an alternate representation for a decimal number.

Let me show you what I mean.

The number 0.7 can be written in many different ways:

$$0.7$$
$$= 0.70$$
$$= 0.700$$
$$= 0.7000$$
$$= 0.70000$$
etc.

This is true since $0.70 = \frac{70}{100}$, which reduces to $\frac{7}{10}$; and $.700 = \frac{700}{1000}$, which also reduces to $\frac{7}{10}$. Similarly all the other representations reduce to $\frac{7}{10}$.

Those zeros that are attached after the decimal point do not change the value of the decimal number; they simply give other equivalent numbers. By attaching zeros, we likewise get:

$$\begin{array}{ll} 52 & 1.89 \\ = 52. & = 1.890 \\ = 52.0 & = 1.8900 \\ = 52.00 & = 1.89000 \\ = 52.000 & = 1.890000 \\ \text{etc.} & \text{etc.} \end{array}$$

Example 1

$.5762 - .34 = ?$

Line up the decimal points.

$$\begin{array}{r} .5762 \\ -.3400 \\ \hline .2362 \end{array}$$

Change the .34 to .3400 to get the same number of digits after the decimal point.

Example 2

$56.4 - 3.27 = ?$

$$\begin{array}{r} 56.40 \\ -\ 3.27 \\ \hline 53.13 \end{array}$$

Change the 56.4 to 56.40 to get the same number of digits after the decimal point.

Example 3

$$189 - 13.567 = ?$$

$$\begin{array}{r} 189.000 \\ -\ 13.567 \\ \hline 175.433 \end{array}$$ Change the 189 to 189.000 to get the same number of digits after the decimal point.

This then is a summary of subtracting decimal numbers.

SUBTRACTING DECIMALS

1. Get the same number of digits after the decimal point in both numbers by attaching zeros after the decimal point.
2. Line up the decimal points of both numbers.
3. Subtract as usual.
4. Place the decimal point in the answer below the other decimal points.

PROBLEMS

1) $23.62 - 19.048 = ?$ *Answer*: 4.572

$$\begin{array}{r} 23.620 \\ -\ 19.048 \\ \hline 4.572 \end{array}$$

2) $57.2457 - 20.7 = ?$ *Answer*: 36.5457

$$\begin{array}{r} 57.2457 \\ -\ 20.7000 \\ \hline 36.5457 \end{array}$$

3) $901 - 23.452 = ?$ *Answer*: 877.548

$$\begin{array}{r} 901.000 \\ -\ 23.452 \\ \hline 877.548 \end{array}$$

To understand how to multiply decimal numbers, we must look at the fractions which the decimal numbers represent.

For example: $0.3 \times 0.71 = ?$

$$0.3 = \frac{3}{10} \text{ and } 0.71 = \frac{71}{100}$$

So, $0.3 \times 0.71 = \dfrac{3}{10} \times \dfrac{71}{100}$

$$= \frac{213}{1000}$$

$$= 0.213$$

Even rabbits can multiply quickly.

If every time you wanted to multiply decimal numbers you had to change them to fractions, it could become a very involved process. So as we have done previously, we will use that example to obtain a shorter way to multiply decimals.

Let us look at that example again.

$$0.3 \times 0.71 = 0.213$$

0.3   has 1 decimal digit.
0.71   has 2 decimal digits.
0.213   has 3 decimal digits.

132   CHAPTER 4

> The number of decimal digits in the product is the total of the number of decimal digits in the numbers being multiplied.

We can now do the above example as follows:
Write the numbers above each other, ignoring the decimal points, and multiply the numbers as if they were whole numbers.

```
      0.71        has 2 decimal digits.
    × 0.3         has 1 decimal digit.
     .213         The answer has a total of 3 decimal digits.
```

Example 1

$15.31 \times 0.07 = ?$

```
     15.31        has 2 decimal digits.
   × 0.07         has 2 decimal digits.
    1.0717        The answer has a total of 4 decimal digits.
```

Example 2

$19.86 \times 0.089 = ?$

```
     1 9.8 6      has 2 decimal digits.
   × 0.0 8 9      has 3 decimal digits.
     1 7 8 7 4
     1 5 8 8 8
     1.7 6 7 5 4  The answer has a total of 5 decimal digits.
```
(*Note:* To place the decimal in the answer, start from the right and move 5 places to the left.)

Example 3

$0.0002 \times 0.007 = ?$

```
     0.0 0 0 2    has 4 decimal digits.
   × 0.0 0 7      has 3 decimal digits.
    .0 0 0 0 0 1 4   The answer has a total of 7 decimal digits.
```
(*Note:* 5 zeros are used as place holders in front of the 14 to get the 7 decimal digits.)

PROBLEMS

1) $43.6 \times 2.7 = ?$ *Answer:* 117.72

```
        4 3.6
      ×  2.7
        3 0 5 2
        8 7 2
      1 1 7.7 2
```

2) $6.075 \times 3.14 = ?$ *Answer:* 19.0755

```
        6.0 7 5
      × 3.1 4
        2 4 3 0 0
        6 0 7 5
      1 8 2 2 5
      1 9.0 7 5 5 0
```

3) $5.3 \times 0.006 = ?$ *Answer:* 0.0318

$$\begin{array}{r} 5.3 \\ \times\, 0.006 \\ \hline .0318 \end{array}$$

4) What is wrong with the solution below?

$$\begin{array}{r} 43 \\ \times\, .05 \\ \hline 21.5 \end{array}$$

Answer:
The decimal point should be placed two digits from the *right* of the number. The answer is 2.15.

Dividing decimals is done using similar methods as used in dividing whole numbers. The only difference is in placing the decimal point in the answer. *If the divisor is a whole number, the decimal point is simply placed in the answer above its position in the dividend and the division is done ignoring the decimal point.* The reasoning here is that the divisor is dividing into the whole number part and the fractional part with the decimal point used to separate both parts in the answer.

Example 1

$52.8 \div 4 = ?$

Place the decimal point above the decimal point in the dividend. That separates the whole and fractional parts of the answer.

whole number divisor →
$$\begin{array}{r} 13.2 \\ 4\overline{)52.8} \\ \underline{4} \\ 12 \\ \underline{12} \\ 8 \\ \underline{8} \end{array}$$

If the divisor is not a whole number, you can make it a whole number by moving its decimal point to the right. You must, however, also move the decimal point the same number of places to the right in the dividend.

Example 2

$2.38 \div .7 = ?$

$$\begin{array}{r} 3.4 \\ .7\overline{)2.3.8} \\ \underline{2\,1} \\ 2\,8 \\ \underline{2\,8} \end{array}$$

1. Move the decimal point to change the divisor into a whole number.
2. Move the point the same in the dividend.
3. Place the decimal point in the answer above the moved decimal point.
4. Do the division ignoring the decimal points.

What you are really doing when you move the decimal point one place to the right is multiplying the divisor and the dividend by 10.

$$2.38 \div .7 = \frac{2.38}{.7} \xrightarrow{\times 10} \frac{23.8}{7} = 23.8 \div 7$$

You have transformed the original problem into an equivalent problem, using the same methods as raising fractions to higher terms. Similar logic can be used to justify movement of the decimal point in other division problems.

Example 3

1.30647 ÷ 4.07 = ?

```
           .3 2 1
    4.0 7)1.3 0 6 4 7
          1 2 2 1
            8 5 4
            8 1 4
              4 0 7
              4 0 7
```

1. Move the decimal point to change the divisor into a whole number.
2. Move the point the same in the dividend.
3. Place the decimal point in the answer above the moved decimal point.
4. Do the division ignoring the decimal points.

Example 4

21 ÷ .0028 = ?

```
            7 5 0 0.
    .0 0 2 8)2 1.0 0 0 0
             1 9 6
               1 4 0
               1 4 0
                   0
                   0
                     0
                     0
```

1. Move the decimal point to change the divisor into a whole number.
2. Move the point the same in the dividend. (Notice that four zeros were attached.)
3. Place the decimal point in the answer above the moved decimal point.
4. Do the division ignoring the decimal points.

Example 5

Find the exact answer for .23 ÷ 18.4.

(Finding the exact answer means to continue dividing until there is no remainder. The answer is not to be rounded off.)

```
              .0 1 2 5
    1 8 .4).2 3 0 0 0
            1 8 4
              4 6 0
              3 6 8
                9 2 0
                9 2 0
```

Zeros were attached one at a time until there was no remainder in the division.

To find the exact answer in a division problem, it may be necessary to attach zeros after the decimal point as we did in the last example.

PROBLEM

1) 510.17 ÷ 6.002 = ?   *Answer:* 85

```
                8 5.
    6.0 0 2)5 1 0.1 7 0
            4 8 0 1 6
              3 0 0 1 0
              3 0 0 1 0
```

Name _____ Date _____

EXERCISE 4.3

Find each answer.

1) 145.21 + 93.66 + 272.54

2) 102.3 + 20.03 + 2.303

3) 0.34 + 0.99 + 0.01 + 0.66

4) 10.5 + 3.89 + 5 + 21.7

5) 345.67 − 123.51

6) 839.99 − 253.99

7) 600 − 230.55

8) 8.5 − 2.03

9) 451.69 − 25

10) 2345.11 − 963.33

11) 745(0.8)

12) 9.5(2.1)

13) 32.05(4.6)

14) 1.34(2.55)

15) 3245.67(10)

16) 3245.67(100)

17) 3245.67(1000)

18) 3245.67(10000)

19) 235.45 ÷ 5

20) 8674 ÷ 0.2

21) 1362.21 ÷ 0.03

22) 912.75 ÷ 1.5

23) 2145.0671 ÷ 10

24) 2145.0671 ÷ 100

25) 2145.0671 ÷ 1000

26) 2145.0671 ÷ 10000

27) Find the sum of 0.3, 0.9, and 0.27.

28) Find the difference of 5000 and 3589.99.

29) Find the product of 2, 2.2, and 2.02.

30) Find the quotient of 30.28 and 0.04

4.4 CONVERTING FRACTIONS AND DECIMALS

CHANGING FRACTIONS TO DECIMALS

To convert a fraction to a decimal you must remember that the fraction line indicates a division—divide the numerator by the denominator.

For example, $\frac{3}{4} = 3 \div 4$

$$\begin{array}{r} .75 \\ 4\overline{)3.00} \\ \underline{2\ 8} \\ 20 \\ \underline{20} \end{array}$$

Thus, $\frac{3}{4}$ is the decimal 0.75.

Example 1

Express $\frac{7}{12}$ as a decimal rounded to two decimal places.

$$\frac{7}{12} = 12\overline{)7.000} \approx 0.58$$

with divisor 12, partial results 60, 100, 96, 40, 36.

Example 2

Express $2\frac{3}{5}$ in decimal form.

$$2\frac{3}{5} = \frac{13}{5} = 5\overline{)13.0} = 2.6$$

with divisor 5, partial results 10, 30, 30.

Commonly used fraction-decimal equivalents are obtained by using the division process above and are given in the chart below.

FRACTION-DECIMAL EQUIVALENTS

$\frac{1}{4} = 0.25$	$\frac{1}{2} = 0.5$	$\frac{3}{4} = 0.75$	
$\frac{1}{3} \approx 0.33$	$\frac{2}{3} \approx 0.67$		
$\frac{1}{5} = 0.2$	$\frac{2}{5} = 0.4$	$\frac{3}{5} = 0.6$	$\frac{4}{5} = 0.8$
$\frac{1}{8} = 0.125$	$\frac{3}{8} = 0.375$	$\frac{5}{8} = 0.625$	$\frac{7}{8} = 0.875$

CHANGING DECIMALS TO FRACTIONS

To convert a decimal number to a fraction, use the place values of digits to the right of the decimal point—tenths, hundredths, thousandths, ten-thousands, etc. Using those place values and the way decimals numbers are read, decimals can be converted to fractions.

For example: 0.9 reads "nine tenths": $= \dfrac{9}{10}$

0.35 is read "thirty-five hundredths" $= \dfrac{35}{100} = \dfrac{7}{20}$

0.013 is read "thirteen thousandths" $= \dfrac{13}{1000}$

If there is a whole number to the left of the decimal point, it is a mixed number.

For example: 6.2 is read "six and two tenths" $= 6\dfrac{2}{10} = 6\dfrac{1}{5}$

10.11 is read "ten and eleven hundredths" $= 10\dfrac{11}{100}$

80.9 is read "eighty and nine tenths" $= 80\dfrac{9}{10}$

Example 3

Write 456.012 as a reduced mixed number.

456.012 is read "four hundred fifty-six and twelve thousandths."

$456.012 = 456\dfrac{12}{1000} = 456\dfrac{6}{500} = 456\dfrac{3}{250}$

PROBLEMS

1) Write $4\dfrac{3}{17}$ as a two place decimal.   *Answer*: 4.18

$$4\dfrac{3}{17} = \dfrac{71}{17} = 17\overline{)71.000}^{\,4.176} \approx 4.18$$

Write as reduced fractions or mixed numbers: *Answers*:

2) 0.32 $\dfrac{32}{100} = \dfrac{16}{50} = \dfrac{8}{25}$

3) 5.375 $5\dfrac{375}{1000} = 5\dfrac{75}{200} = 5\dfrac{3}{8}$

4) 76.8 $76\dfrac{8}{10} = 76\dfrac{4}{5}$

EXERCISE 4.4 SET A

In problems 1–6, write each fraction as an exact decimal.

_____ 1) $\dfrac{3}{5}$ _____ 2) $\dfrac{7}{14}$

_____ 3) $\dfrac{5}{8}$ _____ 4) $\dfrac{1}{16}$

_____ 5) $\dfrac{41}{32}$ _____ 6) $\dfrac{37}{32}$

In problems 7–12, write each fraction as a three digit decimal.

_____ 7) $\dfrac{6}{11}$ _____ 8) $\dfrac{5}{6}$

_____ 9) $1\dfrac{6}{7}$ _____ 10) $2\dfrac{4}{9}$

_____ 11) $23\dfrac{2}{3}$ _____ 12) $37\dfrac{7}{9}$

In problems 13–20, write each as a reduced fraction or mixed number.

_____ 13) 0.6 _____ 14) 0.4

_____ 15) 0.62 _____ 16) 0.74

_____ 17) 0.0425 _____ 18) 0.0475

_____ 19) 4.001 _____ 20) 3.009

EXERCISE 4.4 SET B

In problems 1–6, write each fraction as an exact decimal.

_____ 1) $\dfrac{4}{5}$ _____ 2) $\dfrac{9}{12}$

_____ 3) $\dfrac{3}{8}$ _____ 4) $\dfrac{5}{16}$

_____ 5) $\dfrac{39}{32}$ _____ 6) $\dfrac{35}{32}$

In problems 7–12, write each fraction as a three digit decimal.

_____ 7) $\dfrac{5}{9}$ _____ 8) $\dfrac{4}{7}$

_____ 9) $2\dfrac{3}{13}$ _____ 10) $3\dfrac{5}{11}$

_____ 11) $44\dfrac{1}{6}$ _____ 12) $32\dfrac{1}{7}$

In problems 13–20, write each as a reduced fraction or mixed number.

_____ 13) 0.2 _____ 14) 0.6

_____ 15) 0.86 _____ 16) 0.98

_____ 17) 0.0375 _____ 18) 0.0275

_____ 19) 1.007 _____ 20) 2.003

4.5 APPLICATIONS WITH DECIMALS

The numbers in the word problems of this section are decimals. However, the problem solving model given in a previous section still applies in solving the word problems.

> **POLYA MODEL: 4 STAGES IN PROBLEM SOLVING**
>
> 1. Read the problem, more than once if needed. As you read the problem, ask yourself what is the given information and is that all the facts.
> 2. Devise a plan or strategy to determine the operations needed to solve the problem.
> 3. Solve the problem. Do the computation from your plan/strategy.
> 4. Check your solution. Did you answer the question? Does the answer make sense in relation to the given information?

PRACTICE PROBLEMS

1) James bought a 25-pound bag of mesquite charcoal for $17.25. What was the cost per pound?

We will use the Poyla Model:

Step 1) The given information tells us the total weight of the bag and the total cost.

Step 2) The phrase "cost per pound" asks for the unit price. Therefore, the cost will be divided by the unit (pounds).

Step 3) Do the division: $17.25 ÷ 25 = $0.69

Step 4) When writing the answer include the units. The answer is $0.69 per pound. The answer makes sense as the unit cost should be less than the total cost.

2) Stacey earns $12.50 per hour and time and a half for any hours worked beyond 35 hours. Last week, Stacey worked a total of 42 hours. How much did Stacey earn?

Step 1) The given information tells us the hourly rate for her regular pay for 35 hours, the overtime rate, and the total number of hours worked.

Step 2) The phrase "time and a half" translates into multiplying the regular hourly rate by 1.5. Therefore, we can find the overtime rate and then the total earnings.

Step 3) regular pay: $12.50(35 hours) = $437.50

overtime rate: $12.50(1.5) = $18.75 per hour for any hours over 35 hours

overtime pay: $18.75(7 hours) = $131.25

total earnings: regular pay + overtime pay
$437.50 + $131.25 = $568.75

Step 4) The total earnings is $568.75. In checking the work, we see that we did compute the overtime rate correctly.

EXERCISE 4.5 SET A

DIRECTIONS:

Write answers with money in appropriate form.

Reduce fractions/ratios to lowest terms.

Round decimals to nearest hundredth/cent.

1) Gross earnings is regular pay plus overtime pay. Net pay is gross earnings minus total of deductions. As an accountant for a local candy store, Garth must calculate each employee's net pay for the week.

Employee	Regular Pay	Overtime Pay	Tax Deductions	Insurance Deductions
Jamilah	$235.55	$55.89	$23.45	$43.00
Clara	$287.90	$0	$54.10	$62.50
Nathan	$256.15	$88.00	$35.43	$56.55

   a) What is Jamilah's net pay for this week?
   b) How much more is Nathan's net pay than Clara's net pay?

2) Harriet earns $13.80 per hour and time and a half for any hours worked beyond 35 hours. Last week, Harriet worked a total of 42 hours. How much did Harriet earn?

3) Tori's gross pay last week was $678.99. Her net pay was $599.60. How much tax and insurance deductions was taken out of Tori's pay last week?

4) Frank's regular pay each week is $659.25. How much does Frank earn annually?

5) Each month the business manager for Fits You clothing store budgets $123,456 for salaries, $6784.50 for rent, $1446 for utilities, and $3456 for clothing items. How much does the business manager budget each year for the expenses?

6) Each month Build For You Carpenters pays a total of $245,765.50 in salaries and benefits for its fifty employees. The amount of the benefits totals $97,984.21.
   a) How much does Build For You Carpenters pay in salaries?
   b) If each employee receives the same salary, how much is each employee's salary?

7) In New Jersey, sales tax on items is found by multiplying by 0.07. Iris spent a total of $234.50 on taxable items. How much did Iris pay in sales taxes?

8) Starting January 1st, Tanith's weekly pay increased from $765.50 to $839.50. How much was her hourly rate increase if she works 40 hours a week?

9) Five friends had dinner at the local diner and decided to split the bill evenly. The bill totaled $68.45 and they left a 20% tip in the amount of $14.00. How much did each person pay for their dinner?

10) A long distance phone company charges $1.25 for the first three minutes plus $0.12 for each additional minute of a phone call. In one week, Sadie made several long distance calls to her mom and best friend. On Monday, Sadie spent 45 minutes on the phone with her mom and 20 minutes on the phone with her best friend. On Tuesday, Sadie spent 30 minutes on the phone with her best friend. On Thursday, Sadie spent 20 minutes on the phone with her mom. On Friday, Sadie spent 25 minutes on the phone with her mom and 45 minutes on the phone with her best friend. How much was Sadie's total phone bill for that week?

11) On March 1ˢᵗ Lizzie had a balance of $256.90 in her checking account. On March 2ⁿᵈ Lizzie deposited $345.20 and then wrote a check in the amount of $550. On March 15ᵗʰ Lizzie's paycheck of $1245.66 was deposited directly into her checking account. On March 16ᵗʰ Lizzie wrote checks in the amounts of $265, $987.89, and $45.11. On March 20ᵗʰ, Lizzie's bank charged her account a $4 service fee.

   a) What was Lizzie's balance on March 3ʳᵈ?

   b) What was Lizzie's balance on March 21ˢᵗ?

12) A plumber charges a $75 flat fee for the first hour of labor plus $45.50 for each additional hour of labor. The plumber spent 6 hours fixing a cracked pipe in the sink. How much was Mrs. Hassel charged by the plumber?

13) The Paisley's went on a road trip. At the beginning of the trip, the car odometer read 67,899.5 miles. At the end of the trip, the car odometer read 69,007 miles. If the car used 45 gallons of gasoline, how many miles per gallon did the car get?

14) Prudenza bought a 25-pound bag of fertilizer for $17.25. What was the cost per pound?

15) By comparing the cost per gallon, determine which brand of cooking oil is the better bargain: Brand X at $7.99 for 5 gallons or Brand Y at $12.99 for 8 gallons.

16) John's yearly income is $45,678 and his wife's yearly income is $43,900. What is their combined monthly income?

17) Gary and Elena each smoke 2 packs of cigarettes per day. A pack of cigarettes costs $4.75. How much do Gary and Elena combined spend on cigarettes each year?

18) A 12-ounce can of ready to feed baby formula costs $10.99. Mrs. Gracias uses $3\frac{1}{2}$ cans of formula each day to feed her twin sons. If Mrs. Gracias uses the baby formula for 7 months (210 days), how much did she spend?

19) The Photography Club charges $7.50 for a 5" by 7" print of the Brooklyn Bridge, $5.25 for a 5" by 7" print of the Verrazano Bridge, and $6.50 for a 5" by 7" print of the Manhattan Bridge. At this year's fundraiser, the Club sold 10 Brooklyn Bridge prints, 12 Verrazano Bridge prints, and 5 Manhattan Bridge prints. How much did the Photography Club earn selling the prints?

20) Carol wrote a check for a total of $249.75. She purchased VHS tapes of past winter Olympics at $9.99 each. How many tapes did Carol buy?

21) Nylons and More store bought 5 dozen boxes of navy blue stockings at $2.99 per box. The store sold the stockings at $4.59 per box. How much profit did Nylons and More store make?

22) A map's scale indicates that 1 inch is equal to 5.5 miles. If you measure $3\frac{1}{2}$ inches on the map, then how many miles will that equal?

23) Vera wants to enclose her rectangular garden to keep out the rabbits. The garden measures 8 feet by 9 feet. The local hardware store charges $5.65 per linear foot of fencing. How much did Vera pay for the fence?

24) Lehi has 45 shares of C stock. He has decided to sell $\frac{2}{3}$ of the shares at $7.95 per share. How much money does he make on the sale of the stocks?

25) To make extra money, Eloise sells Christmas wreaths by her own design at $15.50 each. Her goal is to earn at least $1000. How many wreaths would she need to sell to meet her goal?

Name _____   Date _____

EXERCISE 4.5 SET B

DIRECTIONS:

Write answers with money in appropriate form.
Reduce fractions/ratios to lowest terms.
Round decimals to nearest hundredth/cent.

1) Oscar earns $12.50 per hour. If Oscar worked 35.5 hours this week, how much did Oscar earn?

2) Victor earned $393.75 last week. He worked 35 hours. How much did Victor earn per hour last week?

3) Ellen's net pay last month was $1356.95. Her tax and insurance deductions totaled $215.34. What was Ellen's gross pay last month?

4) Michaela earns $3345.50 each month. What is Michaela's annual earnings?

5) Larry's regular pay is $47,820.38 per year. Larry was able to work a great amount of overtime last year and so his annual income was $51,108.75. How much did Larry earn in overtime pay?

6) Delia's gross pay is $2345.39 each month. Her deductions each month are $135.87, $95.67, and $45.66. How much is Delia's net pay?

7) Alice bought 3 bags of potato chips at 75¢ each, 3 bottles of Snapple at $1.25 each, and 3 turkey sandwiches for $3.79 each. How much did Alice spend?

8) The sales tax is found by multiplying the purchases by 0.085. Ulysses bought 2 DVD's for $14.99 each and a calendar for $6.99.

   a) How much was the sales tax?
   b) How much was the total purchase including sales tax?

9) The sales tax is found by multiplying the purchases by 0.055. At Bath Items For You store, Midori bought 3 bottles of lotion at $9.99 each, a bath sponge at $5.40, 2 jars of scented candles at $12.29 each, and a basket at $18.00. How much was her total bill?

10) Rhonda went to a math conference in San Francisco for 6 days. She paid $109.99 per day for a hotel room, $39.99 per day for a rental car, and spent about $75 per day for meals. How much did Rhonda spend on the math conference?

11) On Monday you deposited $435.07 into your checking account which had a balance of $98.56. On Tuesday you wrote checks in the amounts of $105.00, $244.75, and $39.99. On Wednesday you were charged a $7 service fee. What is your balance after Wednesday's transactions?

12) A cellular phone company charges $49.99 per month plus $0.45 per minute for any minutes used over the allowed 400 minutes per month. How much would the monthly bill be if you used a total of 450 minutes in one month?

13) Jasper bought 5.5 pounds of potatoes at $0.99 per pound, 3 pounds of string beans at $1.29 per pound, and 2.3 pounds of tomatoes at $1.99 per pound. If Jasper handed the clerk a $100 bill, how much change did he receive?

14) Domenick paid $12 for a 36-fluid ounce can of motor oil. What was the cost per fluid ounce?

15) By comparing the cost per ounce, determine which brand of laundry detergent is the better bargain: 36.5 ounces of Brand A for $4.65 or 48 ounces of Brand B for $5.50.

16) A tax of $3.07 is charged on each box of headphones purchased by an electronic store. One month the store paid $58.33 in taxes for their headphones. How many boxes of headphones were purchased?

17) You are the payroll supervisor for the Optum Company. Each month the company pays $45,457.75 for employees' salaries and benefits. The benefits are $10,658.90 for health insurance, $4,760.25 for retirement, and $988.00 for life insurance. How much does the company pay each month in salaries?

18) The Yearbook Club bought 250 T-shirts at $12.99 each and then sold all of them for $15.29 each. How much profit did the Yearbook Club make?

19) In December, Peter wrote a check for $4380 to buy 40 top-of-the-line coffee makers. How much did each coffee maker cost?

20) The cost of a pack of gum is 79¢. Emmanuel buys 4 packs of gum each month. How much does Emmanuel spend on gum each year?

21) Roberto bought 12 one-by-fours, each measuring $5\frac{3}{4}$ feet. The cost of the one-by-fours is $1.25 per linear foot. How much did Roberto spend?

22) Mohammed bought wire fencing at $8.99 per linear foot to enclose his business. The rectangular lot measures $60\frac{1}{2}$ feet by $55\frac{3}{4}$ feet. How much did Mohammed pay for the wire fencing?

23) Joy needs to fertilize her 20-acre corn field. The Fertilizer Specialist store sells fertilizer at $13.99 per pound. One acre uses $5\frac{1}{2}$ pounds of fertilizer. How much did Joy pay for the fertilizer?

24) Damien decided to put a $300 down payment on a $1299.99 snow-blower and pay the remaining amount in 6 monthly payments. How much was the monthly payments?

25) Victor wants to buy a used car for $8954. He will put a down payment of $800 and make payments of $485.69 for the next 18 months. How much more is he paying for the car than the selling price?

26) Clara made arrangements to pay for the $5382.75 bedroom set in monthly installments of $358.85. How many months did it take Clara to pay for the bedroom set?

27) The Junior Store pays the assistant manager a salary of $30,789 per year for working 35 hours a week. The Senior Store pays the assistant manager $16.85 per hour for working 35 hours per week. At which store does the assistant manager earn more?

4.6 ORDER OF OPERATIONS WITH FRACTIONS, DECIMALS, AND SIGNED NUMBERS

The order of operations and the rules for operations with signed numbers still hold true for expressions with integers, fractions, and decimals. Return to previous sections and review the rules.

PRACTICE PROBLEMS

1) $8 \div 24 \div 4$ given expression

$\dfrac{8}{24} \div 4$ rewrite as fraction $\left(a \div b = \dfrac{a}{b}\right)$

$\dfrac{1}{3} \cdot \dfrac{1}{4}$ reduce and convert to multiplication

$\dfrac{1}{12}$ multiply

2) $(0.25)^2 - \left(\dfrac{1}{2}\right)^3$ given expression

Approach #1

$\left(\dfrac{1}{4}\right)^2 - \left(\dfrac{1}{2}\right)^3$ convert decimal to fraction

$\dfrac{1}{16} - \dfrac{1}{8}$ simplify exponents

$-\dfrac{1}{16} = -0.0625$ different signs; subtract and keep sign of larger number

Practice Problem 2 can be simplified by another approach; converting the fraction to a decimal.

$(0.25)^2 - \left(\dfrac{1}{2}\right)^3$ given expression

Approach #2

$(0.25)^2 - (0.5)^3$ convert fraction to decimal

$0.0625 - 0.125$ simplify exponents

-0.0625 different signs; subtract and keep sign of larger number

EXERCISE 4.6 SET A

Find each answer.

1) $4 \div 16 \div -2$

2) $9 \div 3 \cdot (6 \div 4 + 2)$

3) $\left(2\dfrac{1}{4}\right)^2 - \dfrac{2}{3} \cdot \dfrac{3}{5}$

4) $(0.4)^2 + 2(5 - 3 \div 5)$

5) $\left(-\dfrac{3}{2}\right)^2 \div \dfrac{5}{6} + 3.4$

6) $7\dfrac{1}{2} - (-5 - 2.32)$

7) $-\dfrac{4}{5} \cdot 12 - (38 \div 5)$

8) $-12.5 + 2^2 + (4 - 4 \div 7)$

9) $\left(-\dfrac{1}{3}\right)^3 \left(\dfrac{1}{2}\right)^4 (6)^2$

10) $(-0.2)^3(-1.0)^3(0.3)^2$

11) $25 - \dfrac{2}{5}(15 \div 26)$

12) $\dfrac{25}{12} \div (1.25)^2 + 3$

13) $100 \div (0.5)^2 \cdot -\dfrac{1}{5}$

14) $\dfrac{1}{2} \cdot (10)^2 \div (0.0004 \cdot 10^4)$

15) $\left(\dfrac{1}{4}\right)(1.5) - 3(4 \div 15)$

16) $\dfrac{1}{2}\left(\dfrac{3}{4} \div \dfrac{1}{8}\right) - \dfrac{1}{3}\left(\dfrac{2}{7} \div \dfrac{1}{14}\right)$

17) $3(0.3) + 0.4(1.2) - 10(2.2)$

18) $\left(2\dfrac{2}{3} \cdot \dfrac{3}{5}\right)^2 - 0.5(6 \div 12)$

19) $0.25 + 1.2 \cdot \dfrac{3}{2} \div 2 - \dfrac{1}{2} \cdot 0.4$

20) $4\dfrac{1}{4}(7 - 4.5) + 3\dfrac{1}{2}(4.5 - 7)$

EXERCISE 4.6 SET B

Find each answer.

1) $9 \div 18 \div -4$

2) $6 \div 3 \cdot (9 \div 5 + 8)$

3) $\left(3\dfrac{1}{4}\right)^2 - \dfrac{8}{5} \cdot -\dfrac{10}{16}$

4) $(0.6)^2 + 3(5 - 4 \div 5)$

5) $\left(-\dfrac{4}{3}\right)^2 \div \dfrac{5}{6} - 3.4$

6) $7\dfrac{2}{3} - 3 \cdot (-6 - 8.1)$

7) $-\dfrac{3}{5} \cdot 20 - (28 \div 6)$

8) $-10.4 + 3^2 \cdot (8 - 5 \div 2)$

9) $\left(-\dfrac{2}{3}\right)^3 \left(\dfrac{1}{2}\right)^4 (8)^2$

10) $(-1.0)^3 (-0.3)^3 (0.4)^2$

11) $35 + \dfrac{2}{9}(27 \div 36)$

12) $\dfrac{21}{10} \div (1.75)^2 - 3$

13) $200 \div (0.4)^2 \cdot -\dfrac{1}{4}$

14) $\dfrac{1}{5} \cdot (10)^2 \div (0.0005 \cdot 10^4)$

15) $\left(\dfrac{1}{2}\right)(2.5)^2 - 3(4 \div 15)$

16) $\dfrac{1}{2}\left(\dfrac{2}{7} \div \dfrac{1}{8}\right) - \dfrac{1}{3}\left(\dfrac{2}{3} \div \dfrac{1}{4}\right)$

17) $4(0.4) + 0.6(1.2) - 100(.42)$

18) $\left(\dfrac{2}{5} \cdot \dfrac{3}{2}\right)^2 - 0.3(10 \div 12)$

19) $0.75 + 2.2 \cdot \dfrac{5}{4} \div 4 - \dfrac{1}{2} \cdot 0.2$

20) $1\dfrac{1}{3}(8 - 0.5) + 2\dfrac{1}{2}(0.5 - 8)$

5.1 RATIOS

Ratio:
After Anna's song, get ready for the laser demonstration. It's an awesome ray show!

If there were 18 males and 12 females in a class, you could compare the number of men to women by saying there is a ratio of 18 men to 12 women. You could represent that comparison in three different ways:

$$18 \text{ to } 12$$
$$18 : 12$$
$$\frac{18}{12}$$

The ratio of 18 to 12 is another way to represent the fraction $\frac{18}{12}$. The three representations above are equal.

$$18 \text{ to } 12 = 18 : 12 = \frac{18}{12}$$

Depending on the situation, any of those three forms can be used. *A ratio, then, is simply a comparison of numbers that can be expressed as a fraction.*

The first operation to perform on a ratio is to reduce it to lowest terms. The above ratio, 18 : 12, can be reduced to lower terms just as we reduced fractions.

$$18 : 12 = \frac{18}{12} \xrightarrow{\div 6} \frac{3}{2}$$

$$18 : 12 = \frac{3}{2} = 3 : 2$$

By reducing the ratio you can get a better understanding of the original ratio. The ratio of 18 males to 12 females reduces to a ratio of 3 males to 2 females. This means that for every 3 males in the class there are 2 females.

Example 1

Express the ratio of the value of 4 dimes to 3 quarters.

You must first find the value of each by changing them to the same unit (cents).

$$4 \text{ dimes} = 40¢ \text{ and } 3 \text{ quarters} = 75¢$$

Thus, the ratio of the respective values is:

$$40 : 75 = \frac{40}{75} = \frac{8}{15}$$

1.) Pages 009–018, 037–042, 045–047, 052–059, 062–077, 080–087, 115–127, 129–132, 141–143, 147–152, 184–187, 199–203, 206–209, from *Math for College Students: Arithmetic with Introductions to Algebra and Geometry*, 5th Edition by Ronal Straszkow. © 1997 by Kendall/Hunt Publishing Company. Used with permission.

Example 2

A basketball team wins 16 games and loses 14 games. Find the reduced ratio of:

$$\text{wins to losses.} \qquad 16 : 14 = \frac{16}{14} = \frac{8}{7}$$

$$\text{losses to wins.} \qquad 14 : 16 = \frac{14}{16} = \frac{7}{8}$$

$$\text{wins to total games played.} \qquad 16 : 30 = \frac{16}{30} = \frac{8}{15}$$

Notice that the order of the numbers is critical.

Example 3

Find the reduced ratio of $5\frac{2}{3}$ to 2.5.

$$5\frac{2}{3} : 2.5 = \frac{5\frac{2}{3}}{2.5} = \frac{5\frac{2}{3}}{2\frac{1}{2}}$$

$$= \frac{\frac{17}{3}}{\frac{5}{2}} = \frac{17}{3} \div \frac{5}{2}$$

$$= \frac{17}{3} \times \frac{2}{5} = \frac{34}{15}$$

If you keep in mind that a ratio is simply a comparison of quantities expressible as a fraction, you should be able to do the following problems.

PROBLEM

1) What is the reduced ratio $3\frac{1}{4}$ inches to 3 feet?   *Answer:* 13:144

(*Note:* 3 ft = 36 in.)

$$3\frac{1}{4} : 36 = \frac{3\frac{1}{4}}{36} = 3\frac{1}{4} \div 36$$

$$= \frac{13}{4} \div \frac{36}{1}$$

$$= \frac{13}{4} \times \frac{1}{36} = \frac{13}{144}$$

EXERCISE 5.1 SET A

A team wins 8 games out of 12 games played. What is the reduced ratio of the following:

_____ 1) wins to games played?

_____ 2) wins to losses?

_____ 3) losses to games played?

A jar contains 12 white, 10 red, and 18 blue balls. What is the reduced ratio of the following:

_____ 4) white balls to blue balls?

_____ 5) red balls to the total number of balls?

_____ 6) blue balls to balls that are not blue?

A bank contains 7 dimes, 20 nickels, and 9 quarters. What is the reduced ratio of the value of the following:

_____ 7) nickels to dimes?

_____ 8) dimes to nickels?

_____ 9) dimes to quarters?

_____ 10) quarters to total money in the bank?

In problems 11–17, what is the reduced ratio of the distances:

_____ 11) 18 inches to 45 inches?

_____ 12) 5 inches to 5 feet?

_____ 13) 8 feet to 10 yards?

_____ 14) $11\frac{3}{8}$ inches to $8\frac{1}{2}$ inches?

_____ 15) $3\frac{3}{4}$ inches to $10\frac{1}{2}$ inches?

_____ 16) 9 feet to $3\frac{3}{8}$ inches?

_____ 17) $20\frac{1}{4}$ inches to 3 yards?

EXERCISE 5.1 SET B

DIRECTIONS:

Write answers with money in appropriate form.
Reduce fractions/ratios to lowest terms.
Round decimals to nearest hundredth/cent.

1) A company with 240 employees has the following demographics.

	18–25 years old	26–40 years old	41–62 years old	62 years and older
Male	4	48	36	39
Females	4	56	21	32

Write each ratio as a fraction in simplest form.
   a) male employees to total employees
   b) female employees to total employees
   c) 18–25 year old female employees to total female employees
   d) 41–62 year old male employees to 41–62 female employees

2) In Randy's piggy bank, there were five $20 bills, eight $10 bills, twelve $5 bills and forty-four $1 bills. Write each ratio as a fraction in simplest form of the following:
   a) value of $20 bills to value of $5 bills
   b) value of $20 bills and $10 bills combined to $5 bills and $1 bills combined
   c) value of $10 bills to total value in piggy bank

For #'s 3–16: Write each ratio as a fraction in simplest form.

3) 25 inches to 135 inches

4) $4\frac{3}{5}$ feet to $8\frac{1}{2}$ feet

5) 20.5 pounds to 32.45 pounds

6) 9.3 grams to $\frac{10}{3}$ grams

7) 16 inches to 5 feet

8) $7\frac{1}{2}$ feet to 38 inches

9) 12 yards to 9.5 feet

10) 32.4 feet to $10\frac{1}{4}$ yards

11) $8\frac{1}{4}$ pounds to $8\frac{1}{4}$ ounces

12) 1800 pounds to 3 tons

13) 250 days to 1 year

14) 3 years to 120 weeks

15) $2\frac{1}{2}$ years to 30 months

16) 100 seconds to 10 minutes

17) Write each rate as a unit cost (cost per item).
   a) 55 pounds for $39.16
   b) $28 for 12 gallons
   c) $4\frac{1}{2}$ yards for $18.50

18) Write each rate as a unit rate.
   a) 350 miles in 6 hours (miles per hour)
   b) 45 gallons for 180 miles (miles per gallon)
   c) 6.5 feet in 13 minutes (feet per minute)

19) Desiree bought 5 feet of cotton fabric for $8.75. How much was the cost per foot?

20) Each week Bobby collects 450 plastic bottles from 3 garbage dumpsters in his apartment complex. Cindy collects 1000 plastic bottles from 6 garage dumpsters. Is the rate of plastic bottles to dumpster the same for both people?

21) The Buy Food Supermarket discards about 200 pounds of rotting vegetables and fruits each week. The Fresh Food Supermarket discards about 10,000 pounds of rotting vegetables and fruits each year. Which supermarket has the greater rate of discarded vegetables and fruits per day?

22) In a survey of 200,000 people, 135,000 of them indicated they would like to have a new football stadium built in their city. Write the ratio of the number of people who would like the stadium to the number of people surveyed.

23) Ten out of fifteen smokers surveyed want to stop smoking. Find the ratio of smokers who do not want to stop to the smokers surveyed.

24) The high school basketball team's ratio of wins to losses (no ties) for last season was 24 to 18. Find the ratio of wins to games played last season.

25) Jerry sold 20 boosters for his yearbook at $8.00 each. Helen sold 35 boosters for her yearbook at the same price. Find the ratio of Jerry's total value in boosters sold to Helen's total value in booster's sold.

26) In Lilith's sculpting class there are twelve women and six men. Find the ratio of men in the class to the total number of people in the class.

27) A recent survey questioned if people would travel to a foreign country. The following were the responses; 135 people said yes, 104 people said no, and 26 people said maybe. Find the ratio of people who said yes to the total number of people surveyed.

28) On Monday, Harold decided to sell only two flavors of ice cream cones. He sold 235 chocolate ice cream cones and 355 vanilla ice cream cones. Find the ratio of chocolate ice cream cones to total number of cones sold on Monday.

29) The shoe store had 120 pairs of sneakers in stock on Monday. A shipment of sneakers was delivered to the store on Tuesday and there are now 210 pairs of sneakers in stock. Find the ratio of the number of pairs of sneakers in the shipment to the number of pairs of sneakers in stock on Tuesday.

30) Bob's inventory of cans of soup showed that there were 12 cans of cream of mushroom soup, 15 cans of chicken soup, 8 cans of beef barley, and 12 cans of clam chowder. Each can of soup costs $0.95. Find the ratio of the value of beef barley cans to the total value of the soup cans.

31) A shipment of books has increased the value of inventory from $120,000 to $185,000. Find the ratio of the value of the inventory's increase to the value before the shipment.

32) On Monday, a sports clothing store had 85 Oakland Raiders sweatshirts in stock. On Friday, the store had 50 sweatshirts in stock. Find the ratio of the decrease in Raiders sweatshirts to the sweatshirts in stock on Monday.

33) A landscaper had 100 5-pound bags of fertilizer. In the past week, he had 4 landscaping jobs in which he used 8 bags of fertilizer for each job. Find the ratio of bags of fertilizer used to the number of bags of fertilizer left.

34) Two clothing companies have published their data on sales, costs and profits for the month.

	Clothing Company 1	Clothing Company 2
Sales	$185,050	$202,140
Costs	$95,000	$102,500
Profits	$90,050	$99,640

a) Find the ratio of sales to costs for each company.
b) Is the ratio of profits to costs the same for both companies?
c) Which company has the greater profit to sales ratio?

35) The table below reflects the number of novels sold last week in each category from three bookstores. In each bookstore, the cost of a mystery novel is $6.99; a romance novel is $7.99; and a science fiction novel is $7.49.

	Bookstore A	Bookstore B	Bookstore C
Mystery Novels	144	155	134
Romance Novels	168	120	130
Science Fiction Novels	100	96	188

a) Which bookstore had the greatest ratio of mystery novels sold to science fiction novels sold?
b) Find the ratio of romance novels sold to total novels sold at bookstore C.
c) Find the ratio of the total sales of science fiction novels in Bookstore A to total sales of romance novels in Bookstore A.

5.2 PROPORTIONS

A **proportion** is a statement that one ratio is equal to another ratio.
For example, a ratio of 4 : 8 is equal to a ratio of 3 : 6.

$$4 : 8 = \frac{4}{8} = \frac{1}{2} \text{ and } 3:6 = \frac{3}{6} = \frac{1}{2}$$

$$4 : 8 = 3 : 6$$

$$\frac{4}{8} = \frac{3}{6}$$

Those ratios form a proportion since they are equal to each other. In a proportion, you will notice that if you **cross multiply** the terms of a proportion as shown below, those **cross-products** are equal.

$$\frac{4}{8} = \frac{3}{6} \qquad 4 \times 6 = 8 \times 3 \text{ (both equal 24)}$$

$$\frac{3}{2} = \frac{18}{12} \qquad 3 \times 12 = 2 \times 18 \text{ (both equal 36)}$$

This, then, is the **fundamental principle** for working with proportions:

> If you cross multiply in a proportion both answers are equal

Example 1

Do $\frac{4}{7}$ and $\frac{36}{63}$ form a proportion?

Cross multiply the ratios. $\quad \frac{4}{7} = \frac{36}{63} \quad \begin{array}{l} 4 \times 63 = 252 \\ 7 \times 36 = 252 \end{array}$

The cross-products are equal, so the ratios form a proportion.

That fundamental principle of proportions enables you to solve problems in which one number of the proportion is not known. For example, if N represents the number that is unknown in the proportion below, we can find its value.

$$\frac{N}{12} = \frac{3}{4}$$

$4 \times N = 12 \times 3$ Cross multiply the proportion.

$4 \times N = 36$ We know that N = 9 since $4 \times 9 = 36$. However, other problems may not be so easy. You should learn this method for finding the value for N:

$\dfrac{4 \times N}{4} = \dfrac{36}{4}$ *Divide the terms on both sides of the equals sign by the number next to the unknown letter.*

$1 \times N = 9$ In this problem, we would divide both sides by 4.

$N = 9$ That will leave the N on the left side and the answer (9) on the right side.

We covered the basic steps for solving a proportion. By using those methods and your knowledge of fractions and decimals, you can now attempt problems that are more difficult.

Example 1

Do $1\frac{3}{4} : \frac{2}{5}$ and $2\frac{11}{12} : \frac{2}{3}$ form a proportion?

Proportion: The N. F. L. scores are in the pro-portion of the radio sports program.

$$\frac{1\frac{3}{4}}{\frac{2}{5}} \stackrel{?}{=} \frac{2\frac{11}{12}}{\frac{2}{3}}$$

$$1\frac{3}{4} \times \frac{2}{3} \stackrel{?}{=} \frac{2}{5} \times 2\frac{11}{12} \qquad \text{Cross multiply.}$$

$$\frac{7}{\underset{2}{\cancel{4}}} \times \frac{\cancel{2}^1}{3} \stackrel{?}{=} \frac{\cancel{2}^1}{\cancel{5}_1} \times \frac{\cancel{35}^7}{\cancel{12}_6}$$

$$\frac{7}{6} = \frac{7}{6}$$

Yes, they do form a proportion.

Example 2

Solve for P. $\dfrac{2}{3} = \dfrac{P}{100}$

$3 \times P = 200$ Cross multiply.

$\dfrac{3 \times P}{3} = \dfrac{200}{3}$ Divide by the number next to the unknown letter; divide by 3.

$P \approx 66.67$

Example 3

Solve for N. $\dfrac{4.6}{3.75} = \dfrac{N}{7.875}$

$3.75 \times N = 4.6 \times 7.875$ Cross multiply.

$\dfrac{3.75 \times N}{3.75} = \dfrac{36.225}{3.75}$ Divide by 3.75.

$N = 9.66$

Example 4

Solve for B. $\dfrac{\frac{1}{2}}{B} = \dfrac{\frac{1}{4}}{100}$

$\dfrac{1}{4} \times B = \dfrac{1}{2} \times 100$ Cross multiply.

$\dfrac{\frac{1}{4} \times B}{\frac{1}{4}} = \dfrac{50}{\frac{1}{4}}$ Divide by $\dfrac{1}{4}$.

$B = 50 \div \dfrac{1}{4}$
$B = 50 \times 4$
$B = 200$

PROBLEMS

1) Do 16.4 : 5.6 and 18.4 : 7.6 form a proportion?

   Answer: No.
   The cross products are not equal.

   $\dfrac{16.4}{5.6} \underset{\longleftarrow}{\overset{\longrightarrow}{=}} \dfrac{18.4}{7.6}$

   $16.4 \times 7.6 = 124.64$
   $5.6 \times 18.4 = 103.04$

2) Solve for A. $\dfrac{A}{\frac{17}{3}} = \dfrac{21}{100}$

   Answer: A = 1.19

   $100 \times A = \dfrac{17}{\cancel{3}} \times \overset{7}{\cancel{21}}$

   $\dfrac{100 \times A}{100} = \dfrac{119}{100}$
   $A = 1.19$

3) Solve for B. $\dfrac{1.6}{B} = \dfrac{3.5}{6.39}$

   Answer: B ≈ 2.92

   $3.5 \times B = 1.6 \times 6.39$
   $\dfrac{3.5 \times B}{3.5} = \dfrac{10.224}{3.5}$
   $B \approx 2.92$

Name _____ Date _____

EXERCISE 5.2

In problems 1–4, determine if the ratios form a proportion.

_____ 1) $2\frac{2}{3} : 1\frac{1}{5}$ and $1\frac{2}{3} : \frac{3}{4}$

_____ 2) $1\frac{1}{4} : 3\frac{1}{3}$ and $1\frac{1}{2} : 4$

_____ 3) 1.4 to 5.6 and 0.4 to 1.6

_____ 4) 0.3 to 3.8 and 0.9 to 11.4

In problems 5–18, find the unknown in the proportion.

_____ 5) $\dfrac{A}{16} = \dfrac{7}{100}$ _____ 6) $\dfrac{A}{15} = \dfrac{9}{100}$

_____ 7) $\dfrac{5}{8} = \dfrac{7}{N}$ _____ 8) $\dfrac{4}{9} = \dfrac{7}{N}$

_____ 9) $\dfrac{Y}{.35} = \dfrac{1.2}{.17}$ _____ 10) $\dfrac{Y}{.12} = \dfrac{1.8}{.71}$

_____ 11) $\dfrac{5.3}{B} = \dfrac{9}{100}$ _____ 12) $\dfrac{4.6}{B} = \dfrac{7}{100}$

_____ 13) $\dfrac{F}{\frac{1}{2}} = \dfrac{38}{\frac{1}{4}}$ _____ 14) $\dfrac{F}{\frac{1}{4}} = \dfrac{36}{\frac{1}{2}}$

_____ 15) $\dfrac{17}{2.5} = \dfrac{P}{100}$ _____ 16) $\dfrac{28}{3.9} = \dfrac{P}{100}$

_____ 17) $\dfrac{1\frac{3}{4}}{D} = \dfrac{\frac{7}{8}}{1\frac{1}{4}}$ _____ 18) $\dfrac{2\frac{1}{2}}{D} = \dfrac{\frac{5}{6}}{1\frac{1}{6}}$

5.3 APPLICATIONS INVOLVING PROPORTIONS

We discussed how to solve for an unknown quantity in a proportion. In this section, we will use that knowledge to solve real-life problems in which two ratios or rates are equal.

Example 1

At 2 P.M. on a sunny day, a 5 ft woman had a 2 ft shadow, while a church steeple had a 27 ft shadow. Use this information to find the height of the steeple.

Distance:
"May I have dis-dance?"

Analyze: Since both shadows were measured at 2 P.M., the ratio of the height of the woman to her shadow must equal the ratio of the height of the steeple to its shadow. If H represents the height of the steeple, we get the proportion 5 : 2 = H : 27.

Solve: $\dfrac{5}{2} = \dfrac{H}{27}$ $\left[\dfrac{\text{height}}{\text{shadow}} = \dfrac{\text{height}}{\text{shadow}} \right]$

$2 \times H = 5 \times 27$
$2 \times H = 135$
$H = 67.5 \text{ ft}$

Many problems can be solved using proportions to show that two ratios or rates are the same. You must be careful to place the same quantities in corresponding positions in the proportion. In Example 1, we placed the heights of the objects in the numerator of each ratio and the lengths of the shadows in the denominator of each ratio. It would not be logical to let the ratio of a height to its shadow equal the ratio of a shadow to the height of the other object. The quantities in the numerator and denominator of the left of the equal sign must be in the same logical order as the quantities in the ratio on the right of the equal sign.

Example 2

During 15 minutes of television watching, you noticed that there were 2 minutes of commercials. If the amount of commercials continued at the same rate, how many minutes of commercials can you expect in 65 minutes of television watching?

Analyze: Since we are assuming that the commercials will continue at the same rate, we can set up a proportion in which the ratios of commercial time to time watching TV are equal. If N represents the number of minutes of commercials in 65 minutes of TV watching, we get the proportion 2 : 15 = N : 65.

Solve: $\dfrac{2}{15} = \dfrac{N}{65}$ $\left[\dfrac{\text{commercial time}}{\text{time watching TV}} = \dfrac{\text{commercial time}}{\text{time watching TV}} \right]$

$15 \times N = 2 \times 65$
$15 \times N = 130$
$\dfrac{15 \times N}{15} = \dfrac{130}{15}$
$N = 8\dfrac{2}{3} \text{ minutes}$

Example 3
Two hundred tagged fish are released in a small lake. A week later the 300 fish that are caught contain 18 of the tagged fish. Use that data to estimate the number of fish in the lake.

Analyze: The ratio of the number of tagged fish caught to the total tagged fish (18 : 200) should be the same as the ratio of fish caught to the total number of fish (n) in the lake (300 : n).

Solve: $$\frac{18}{200} = \frac{300}{n}$$

$18 \times n = 200 \times 300$
$18 \times n = 60{,}000$
$n = 3333$ (rounded to the ones place)

Example 4
A brass alloy contains only copper and zinc in the ratio of 4 parts of copper to 3 parts zinc. If a total of 140 grams of brass is made, how much copper is used?

Analyze: Since the total amount of alloy (140 g) is known and the amount of copper is unknown, we need to set up a proportion of copper to the total material. The ratio of copper to zinc is 4 : 3, so there are 4 parts copper out of 7 parts of copper and zinc combined. Thus, the ratio of copper to total material is 4 : 7. If N represent the amount of copper in the 140 grams of alloy, the proportion of copper to the total amount is 4 : 7 = N : 140.

Solve: $$\frac{4}{7} = \frac{N}{140} \left[\frac{\text{copper}}{\text{total}} = \frac{\text{copper}}{\text{total}} \right]$$

$7 \times N = 4 \times 140$
$7 \times N = 560$
$N = 80$ grams

PROBLEM

1) At a 4-H Junior Livestock Sale, the advertised hoof-to-freezer ratio for a hog was 220 to 119. That is, if a hog weighed 220 lbs, it should yield 119 lbs of pork after being butchered. At that rate, how many pounds of pork can be expected from a hog that weighs 275 lbs?

Answer: 148.75 pounds

$$\left[\frac{\text{weight}}{\text{yield}} = \frac{\text{weight}}{\text{yield}} \right]$$

$$\frac{220}{119} = \frac{275}{Y}$$

$220 \times Y = 32{,}725$

$Y = 148.75$ lbs

Name _____   Date _____

EXERCISE 5.3 SET A

DIRECTIONS:

Write answers with money in appropriate form.
Reduce fractions/ratios to lowest terms.
Round decimals to nearest hundredth/cent.

1) As inventory manager, Greg documented that for every 5 boxes of cereal sold, 3 boxes of instant oatmeal were sold. If Greg orders 300 boxes of cereal, how many boxes of instant oatmeal should he order?

2) Elementary school nurses know that for every 5 children with complaints of stomach ache, 3 children will vomit. If in one week in the month of January, 45 children visit the nurse's office with a complaint of stomachache, how many of them vomited?

3) A survey of 220 mothers was conducted at a children's department store. The ratio of mothers who preferred candy not be sold at the store to mothers who had no preference was 3 to 2. How many mothers preferred candy not be sold at the store?

4) The owner at a local pizzeria noticed that for every 10 plain pies sold, 7 pies with a topping were sold. A plain pie costs $10.99 and a pie with a topping costs $12.99.
   a) If in one day, 140 pies with a topping were sold, how many plain pies were sold?
   b) What was the total amount of income from both types of pies sold that day?

5) A local entertainment store sells DVD's for $14.99 and video games for $12.29. Records of past sales indicate that for every 8 DVD's sold, 5 video games are sold. In December, the manager expects to sell 400 DVD's. How much income can the manager expect to earn in December in the sales of DVD's and video games?

6) Trevor was charged $5 for every day his taxes were overdue. If Trevor's taxes were overdue for 15 days, how much did Trevor pay in late fees?

7) Jose's commission on selling a $250,000 home was $850. At that rate, how much commission would Jose get on a selling a $600,000 home?

8) Mrs. Tremble sells 3 knitted caps for every 4 knitted scarves. If she sold 450 knitted caps in one year, how many knitted scarves were sold?

9) Federico drinks two 24-ounce bottles of spring water when he works out at the gym for 2 hours. How many ounces of water does Federico drink if in one week he works out a total of $8\frac{1}{2}$ hours?

10) If the scale on a map reads 1 inch = 5.5 miles, then what is the distance between two towns which measures 8 inches on the map?

11) On average, Evan jogs 2 miles in 33.5 minutes. How long will it take Evan if he jogs 7.5 miles?

12) Three 16-ounce bags of seeds are needed to plant 1.5 acres of land for corn. Farmer Ted has 200 acres of land, in which he wants to use only $\frac{3}{4}$ of it to grow corn.
    a) How many acres of land will Farmer Ted use for planting corn?
    b) How many bags of seeds will Farmer Ted need?

13) For every 5 pounds of seedless grapes, $\frac{3}{4}$ pounds are thrown away. A supermarket buys 150 pounds in one week and pays $35.50 for a case of seedless grapes (25 pounds in one case).
    a) How many pounds of grapes will the supermarket throw away?
    b) How much did the supermarket pay for the 150 pounds of grapes?

14) A Beef Stroganoff recipe serving 5 people calls for $1\frac{1}{2}$ cups of cream and $\frac{1}{2}$ cup of milk. Robert will be making Beef Stroganoff to serve 12 people. How many cups of cream and milk does Robert need?

15) The local craft store charges $3.75 for a bag of 24 marbles. Henrietta will make 15 centerpieces for her wedding and will need 36 marbles for each centerpiece.
    a) How many bags of marbles should Henrietta buy?
    b) What will be the total cost for the marbles?
16) If 5.5 pounds of potatoes cost $2.15, how much should 35 pounds of potatoes cost?
17) If one mile equals 5290 yards, how many yards are there in 12 miles?
18) At a park's track, 12 laps is equivalent to 1 mile. Evelyn jogs $2\frac{1}{2}$ miles, 4 days a week. How many laps does Evelyn jog in a year?
19) Johnny, the baker, needs 8 kilograms of chocolate chips to make 256 cookies. He has a 20-pound bag of chocolate chips in his pantry. Is that enough to make all the cookies? *(1 kg = 2.2 lbs)*
20) The local daycare likes to maintain a teacher to child ratio of 1 to 3. If in September the daycare will have 48 children in their care, how many teachers need to be hired?
21) It is estimated that for every 10 eighth-grader, 7 of them smoke at parties. If there are 150,000 eighth-graders in Cook County, how many of them may smoke at parties?
22) For every 10 people who join a gym, 6 of them will go at least three times a week. If 5000 people joined a gym in January, how many of them will go less than three times a week?
23) An intramural soccer team's wins to losses ratio is 3 to 2. In the past two years, the team played 60 games (no ties). How many games did they win?
24) A business manager hires 2 supervisors for every 15 employees. If 24 supervisors were hired, how many employees are in the company?
25) Fifteen thousand pounds of grain are to be divided into two storage areas in the ratio of 5 to 3. How many pounds of grain will be put in each storage area?

EXERCISE 5.3 SET B

DIRECTIONS:

Write answers with money in appropriate form.
Reduce fractions/ratios to lowest terms.
Round decimals to nearest hundredth/cent.

1) Judy, the public relations person for the local theatre, knows that for every 5 people who attend a performance on Friday night, 7 people will attend on Saturday night. Last Friday, 250 people attended the production of Swan Lake. How many could be expected to attend on Saturday night?

2) For every 15 miles that George drives, Randi walks 2 miles. Last week, George drove 150 miles. How many miles did Randi walk?

3) The ratio of men to women in an intramural volleyball league is 4 to 3. Last season, 36 women participated in the league. How many men participated?

4) At a clothing store, the manager orders 12 sweaters for every 5 pairs of jeans. Each sweater costs $9.99 at wholesale value and each pair of jeans costs $15.49 at wholesale value.

    a) If the manager orders 240 sweaters, how many pairs of jeans should she order?

    b) What is the total wholesale cost of the order?

5) At an insurance company, an agent sells 3 home insurance policies for every 14 auto insurance policies. The agent earns a $75 commission for each home insurance policy he sells and a $150 commission for each auto insurance policy he sells. How much total commission did the agent make if he sold 42 auto insurance policies last month?

6) A farmer's tax is $12.50 for every acre of land. Klaus has 250 acres of land. How much does Klaus pay in taxes?

7) Neera makes $85 in commission from selling $300 worth of cosmetics. How much commission does Neera make if she sells $1500 in cosmetics in December?

8) Lisa charged $95.50 for 3 hours of data entry. At that rate, how much will Lisa charge if she does $10 \frac{1}{2}$ hours of data entry?

9) On average, Margo reads 100 pages of a romance novel in $1 \frac{1}{4}$ hours. If she has a block of 3 hours free time, will Margo finish a 350-page romance novel?

10) If the scale on a map reads 1.5 inches = 8 miles, then what is the distance between two cities which measure 12.5 inches on the map?

11) Two 24-ounce bottles of pesticide are needed to spray $8 \frac{1}{2}$ acres of land. Suzanne has 450 acres of land, in which she will only spray the pesticide on 2/3 of the land.

    a) On how many acres of land will Suzanne spray pesticide?

    b) How many bottles of pesticide will Suzanne need to order?

12) The wholesale produce market sells 50 pounds of strawberries for $149.50. If the produce manager of a local supermarket bought 30 pounds of strawberries, how much did he pay?

13) The fabric store has a sale on silk fabric, $18.35 for 5 yards. Judy who loves to make her own silk blouses buys 250 yards. How much does Judy pay for the silk fabric?

14) A Blueberry Crunch pie serves 6 people and needs $3 \frac{1}{4}$ pounds of blueberries. The supermarket sells blueberries at $1.99 per pound. Louise needs to make enough pies to serve 120 people.

    a) How many pounds of blueberries will Louise need to buy?

    b) What will be the total cost of the blueberries?

15) A lamb stew recipe will serve 4 people and requires ___ pounds of boneless lamb shoulder. Anna has invited 10 people to dinner. How many pounds of lamb should Anna buy to make the lamb stew?

16) To make twelve grams of a solution, 2.5 milliliters of carbon are needed. If a lab scientist needs to make 100 grams of the solution, how many milliliters of carbon are needed?

17) At a local pool, 8 laps is equivalent to 1 mile. Michael, a competitor, swims 32 laps, 6 days a week.
   a) How many miles does Michael swim in one week?
   b) How many miles does Michael swim in one year?

18) It takes $1\frac{1}{2}$ minutes to fill a quarter of the gas tank. How many minutes will it take to fill the rest of the gas tank?

19) Garrett spends 3 hours per day watching television for every $\frac{1}{2}$ hour he reads a book or magazine. On average, how many hours will Garrett read in a week if he watched a total of 20 hours during the week?

20) A restaurant likes to maintain a wait staff to table ratio of 2 to 7. If the restaurant has 63 tables, how many wait staff will be needed?

21) It is estimated that for every 8 pregnant women, 5 of them would like to breast feed their child. If 4000 pregnant women are surveyed, then how many of them would like to breast feed their child?

22) In an adult evening class on managing your money, 4 out of every 5 people complete the class. If a total of 250 people are registered for the class, how many of them will not complete the class?

23) In a local 5K marathon, Donald placed 28th out of 200 runners. Donald will enter another 5K marathon in which 500 runners will compete. In what position can Donald expect to place in this race?

24) The parents of a local school district had a vote on whether eighth grade students should be allowed off campus for lunch. For every 2 yes votes, there were 3 no votes. If there were a total of 660 yes votes, how many parents voted?

25) The ratio of men to women in a company is 7 to 6. If there are 390 employees, how many of them are women?

6.1 PERCENTS

A common standard of measurement is the **percent** (%). "Percent" means "out of a hundred." An 85% test score means that out of 100 points you got 85 points. If the sales tax in your state is 7%, it means that for every 100¢ (dollar), there is 7¢ sales tax.

Remember: *"Percent" means "out of a hundred."*
So, 25% means 25 out of 100.

$$25\% = \frac{25}{100} = 0.25$$

137% means 137 out of 100.

$$137\% = \frac{137}{100} = 1.37$$

6.5% means 6.5 out of 100.

$$6.5\% = \frac{6.5}{100} = 0.065$$

Percent:
The odor of a purse.

From those examples, you can see that a percent can be easily expressed as a fraction or a decimal.

CONVERTING PERCENTS TO FRACTIONS
To convert a percent to a fraction, drop the % sign, put the number over 100, and reduce if possible.

CONVERTING PERCENTS TO DECIMALS
To convert a percent to a decimal, drop the % sign and move the decimal point two places to the left.

Example 1

Express 30% as a fraction and as a decimal.

$$30\% = \frac{30}{100} = \frac{3}{10} \text{ (a reduced fraction)}$$

$$30\% = .30 \text{ (a decimal)}$$

Example 2

Express 125% as a mixed number and as a decimal.

$$125\% = \frac{125}{100} = \frac{5}{4} = 1\frac{1}{4} \text{ (a reduced mixed number)}$$

$$125\% = 1.25 \text{ (a decimal)}$$

1.) Pages 009–018, 037–042, 045–047, 052–059, 062–077, 080–087, 115–127, 129–132, 141–143, 147–152, 184–187, 199–203, 206–209, from *Math for College Students: Arithmetic with Introductions to Algebra and Geometry*, 5th Edition by Ronal Straszkow. © 1997 by Kendall/Hunt Publishing Company. Used with permission.

Example 3

Express 9.4% as a decimal.

$$9.4\% = .094$$

Example 4

Express $3\frac{1}{3}\%$ as a fraction.

$$3\frac{1}{3}\% = \frac{3\frac{1}{3}}{100} = 3\frac{1}{3} \div 100$$
$$= \frac{10}{3} \div \frac{100}{1}$$
$$= \frac{\overset{1}{\cancel{10}}}{3} \times \frac{1}{\underset{10}{\cancel{100}}} = \frac{1}{30}$$

You may sometimes find it easier to express a percent as a fraction by converting the percent to a decimal, then converting the decimal to a fraction.

Example 5

Express $18\frac{1}{2}\%$ as a fraction.

Since $\frac{1}{2}$ is easily expressed as the exact decimal .5,

$$18\frac{1}{2}\% = 18.5\% = .185 \text{ (a decimal)}$$
$$.185 = \frac{185}{1000} = \frac{37}{200} \text{ (a reduced fraction)}$$

PROBLEMS

1) Express 5.3% as a decimal and as a fraction.

    Answers: .053 and 53/1000

    decimal: $5.3\% = .053$

    fraction: $0.053 = \frac{53}{1000}$

2) Express $166\frac{2}{3}\%$ as a mixed number.

    Answer: $1\frac{2}{3}$

    $$= 166\frac{2}{3}\% = \frac{166\frac{2}{3}}{100}$$
    $$= 166\frac{2}{3} \div 100 = \frac{500}{3} \div \frac{100}{1}$$
    $$= \frac{\overset{5}{\cancel{500}}}{3} \times \frac{1}{\underset{1}{\cancel{100}}} = \frac{5}{3} = 1\frac{2}{3}$$

Name _____ Date _____

EXERCISE 6.1

In problems 1–12, convert each percent to a reduced fraction or mixed number.

_____ 1) 17%   _____ 2) 23%

_____ 3) 5%   _____ 4) 6%

_____ 5) 8.4%   _____ 6) 6.2%

_____ 7) $9\frac{2}{3}\%$ _____ 8) $7\frac{5}{6}\%$

_____ 9) 136%   _____ 10) 215%

_____ 11) $43\frac{3}{4}\%$ _____ 12) $27\frac{1}{4}\%$

In problems 13–22, convert each percent to a decimal.

_____ 13) 45%   _____ 14) 58%

_____ 15) 5%   _____ 16) 8%

_____ 17) 236%   _____ 18) 189%

_____ 19) 26.5%   _____ 20) 20.4%

_____ 21) $8\frac{1}{4}\%$ _____ 22) $9\frac{3}{4}\%$

6.2 CONVERTING DECIMALS AND FRACTIONS TO PERCENTS

In the previous section, you learned how to change percents to either decimals or fractions. In this section you will learn to do the reverse process, that is, change decimals or fractions to percents.

Remember that "percent" means "out of a hundred." So reading how many hundredths are in a decimal number tells the percent.

$$0.07 = \frac{7}{100} \quad \text{so} \quad 0.07 = 7\%.$$

$$0.9 = .90 = \frac{90}{100} \quad \text{so} \quad 0.9 = 90\%.$$

$$1.23 = 1\frac{23}{100} = \frac{123}{100} \quad \text{so} \quad 1.23 = 123\%.$$

Those examples point out a nice way to convert a decimal to a percent.

CONVERTING DECIMALS TO PERCENTS

To convert a decimal to a percent, move the decimal point two places to the right and attach a % sign.

For example:

$$0.34 = 34\% \qquad 0.01 = 1\%$$
$$0.005 = .5\% \qquad 0.0625 = 6.25\%$$
$$2.75 = 275\% \qquad 1.146 = 114.6\%$$

In Section 4.9, you learned to change a fraction to a decimal by dividing the denominator into the numerator of the fraction. Since the above shows how to change a decimal to a percent, putting the two steps together will convert a fraction to a percent.

CONVERTING FRACTIONS TO PERCENTS

To convert a fraction to a percent, divide the denominator of the fraction into the numerator to get a decimal number, then convert the decimal to a percent.

For example:

$$\frac{3}{4} = 4\overline{)3.00}^{.75} = 75\% \qquad \frac{2}{5} = 5\overline{)2.0}^{.4} = 40\%$$

$$\frac{37}{500} = 500\overline{)37.000}^{.074} = 7.4\% \qquad \frac{8}{12} \approx 12\overline{)8.00000}^{.66666} \approx 66.67\%$$

$$\frac{5}{7} \approx 7\overline{)5.00000}^{.71428} \approx 71.43\% \qquad \frac{13}{11} \approx 11\overline{)13.00000}^{1.18181} \approx 118.18\%$$

174   CHAPTER 6

Since the decimal form of some of those examples had many decimal digits, they were rounded off to the nearest hundredth of a percent. You may likewise have to round off the answers in some of the problems and exercises in this section.

PROBLEMS

1) On a test you got 63 out of 75 possible points. What percent did you get correct?

   Answer: 84%

   $$63 \text{ out of } 75 = \frac{63}{75}$$

   $$\begin{array}{r} .84 \\ 75{\overline{\smash{)}63.00}} \\ \underline{60\ 0} \\ 3\ 00 \\ \underline{3\ 00} \end{array}$$

   $.84 = 84\%$

2) Express a ratio of 5 to 6 as a fraction, as a decimal, and as a percent.

   Answers:

   $$\text{fraction: } 5 \text{ to } 6 = \frac{5}{6}$$

   $$\text{decimal: } 6{\overline{\smash{)}5.00000}}^{\,.83333}$$

   percent: $.83333 \approx 83.33\%$

3) Express 0.063 as a fraction and as a percent.

   Answers:

   $$\text{fraction: } \frac{63}{1000}$$

   percent: $0.063 = 6.3\%$

4) Express $\frac{4}{5}$ as a decimal and as a percent.

   Answers:

   $$\text{decimal: } 5{\overline{\smash{)}4.0}}^{\,.8}$$
   $$\phantom{\text{decimal: } 5)}\underline{4.0}$$

   percent: $.8 = 80\%$

5) Express $2\frac{3}{8}$ as a percent.

   Answer: 237.5%

   $$2\frac{3}{8} = \frac{19}{8} = 8{\overline{\smash{)}19.000}}^{\,2.375}$$

   $2.375 = 237.5\%$

Name _____ Date _____

EXERCISE 6.2 SET A

Determine the missing forms in each problem.

Mixed Number or Fraction	Decimal	Percent
1) $\frac{37}{50}$		
2) $\frac{3}{25}$		
3)	.02	
4)	.08	
5)		5.7%
6)		1.9%
7) $\frac{1}{5}$		
8) $\frac{7}{10}$		
9)	.4675	
10)	.4425	
11) $2\frac{1}{6}$		
12) $1\frac{5}{6}$		
13)	2.375	
14)	1.025	
15)		$6\frac{5}{8}$%

EXERCISE 6.2 SET B

Determine the missing forms in each problem.

Mixed Number or Fraction	Decimal	Percent
1) $\frac{13}{50}$	_____	_____
2) $\frac{2}{25}$	_____	_____
3) _____	.04	_____
4) _____	.06	_____
5) _____	_____	6.1%
6) _____	_____	4.3%
7) $\frac{3}{5}$	_____	_____
8) $\frac{2}{5}$	_____	_____
9) _____	.2375	_____
10) _____	.2425	_____
11) $1\frac{2}{3}$	_____	_____
12) $2\frac{1}{3}$	_____	_____
13) _____	1.125	_____
14) _____	2.075	_____
15) _____	_____	$5\frac{3}{8}\%$

6.3 PERCENTAGE PROBLEMS

Consider this problem again: "On a test you got 63 out of 75 possible points. What percent did you get correct?" Since "percent" means "out of a hundred," we can consider this problem as a proportion: 63 out of 75 is what number out of 100? That is,

$$\frac{63}{75} = \frac{P}{100}$$ (Note: P is used to represent the percent or part out of 100.)

We can get the answer for the percent (P) by solving that proportion.

$$\frac{63}{75} = \frac{P}{100}$$
$$\frac{75 \times P}{75} = \frac{6300}{75}$$
$$P = 84$$

The amount you got correct was 63, the test was based on 75 points, and we discovered that the percent was 84. That relationship can be expressed by the **percent proportion:**

$$\frac{A}{B} = \frac{P}{100}$$

A is the amount
B is the base
P is the percent

The percent proportion can be used to solve different types of percentage problems. If you can identify the amount (A), the base (B), and the percent (P), you can utilize the percent proportion.

Here is how you can identify the **amount (A)**, the **base (B)**, and the **percent (P)**:

1. The percent (P) is written with the word "percent" or the % sign.
2. The base (B) follows the word "of."
3. The amount (A) is the remaining number.

By correctly identifying the percent, base, and amount, you can set up a proportion that will enable you to efficiently solve percentage problems. The examples and problems that follow will show you how this is done.

Example 1

15 is what percent of 50?

1) Identify the A, B, P:

<u>15</u> is <u>what percent</u> of <u>50</u>?
↓ ↓ ↓
A P B
(unknown) (follows "of")

2) Set up the percent proportion:

$$\frac{A}{B} = \frac{P}{100}$$

$$\frac{15}{50} = \frac{P}{100}$$ (Leave the unknown with its letter.)

3) Solve the proportion:
$$\frac{50 \times P}{50} = \frac{1500}{50}$$
$$P = 30$$

NOTE: When using the percent proportion, you do *not* move the decimal point to express the percent. The answer to Example 1 is simply 30%

Example 2

16 is 22% of what number?

1) Identify the A, B, P:

$$\underset{A}{\underline{16}} \text{ is } \underset{P}{\underline{22\%}} \text{ of } \underset{B \text{ (unknown)}}{\underline{\text{what number?}}}$$

2) Set up the percent proportion:

$$\frac{A}{B} = \frac{P}{100}$$

$$\frac{16}{B} = \frac{22}{100}$$

3) Solve the proportion:

$$22 \times B = 16 \times 100$$

$$\frac{22 \times B}{22} = \frac{1600}{22}$$

$$B \approx 72.73$$

4) What is 9.5% of 75,000?

Answer: 7125

Percent: $P = 9.5$
Base: $B = 75,000$
Amount: A unknown

$$\frac{A}{75,000} = \frac{9.5}{100}$$

$$\frac{100 \times A}{100} = \frac{712,500}{100}$$

$$A = 7125$$

5) $41\frac{1}{4}$ is what percent of 30?
   137.5

Answer:

Percent: P unknown
Base: $B = 30$
Amount: $A = 41\frac{1}{4} = 41.25$

$$\frac{A}{B} = \frac{P}{100}$$

$$\frac{41.25}{30} = \frac{P}{100}$$

$$\frac{30 \times P}{30} = \frac{4125}{30}$$

$$P = 137.5$$

6) 19.8 is $7\frac{1}{3}\%$ of what number?

Answer: 270

Percent: $P = 7\frac{1}{3}$

Base: B unknown

Amount: $A = 19.8$

$$\frac{A}{B} = \frac{P}{100}$$

$$\frac{19.8}{B} = \frac{7\frac{1}{3}}{100}$$

$$\frac{7\frac{1}{3} \times B}{7\frac{1}{3}} = \frac{1980}{7\frac{1}{3}}$$

$$B = 1980 \div 7\frac{1}{3}$$

$$B = 1980 \div \frac{22}{3}$$

$$B = 1980 \times \frac{3}{22}$$

$$B = 270$$

Name _____   Date _____

EXERCISE 6.3

_____ 1) 9 is what percent of 15?

_____ 2) 28 is what percent of 35?

_____ 3) 20 is 60% of what number?

_____ 4) 18 is 40% of what number?

_____ 5) What is 6% of 50?

_____ 6) What is 5% of 70?

_____ 7) 3.38 is what percent of 13?

_____ 8) 4.2 is what percent of 20?

_____ 9) 18 is 8% of what number?

_____ 10) 15 is 8% of what number?

_____ 11) 32% of 148 is how much?

_____ 12) 43% of 291 is how much?

_____ 13) 63 is what percent of 17.5?

_____ 14) 15 is what percent of 2.5?

_____ 15) 24 is $5\frac{2}{3}\%$ of what number?

_____ 16) 18 is $4\frac{2}{3}\%$ of what number?

_____ 17) What is 17.6% of 45.6?

_____ 18) What is 15.3% of 20.7?

_____ 19) $\frac{3}{4}$ is what percent of $\frac{5}{8}$?

6.4 APPLICATIONS INVOLVING PERCENTS

Percents are used in a variety of real-life problems. In this section, we will investigate other applications of percents.

PERCENT OF A NUMBER

Most percent problems involve either taking a percent of a quantity or using the percent proportion. You will, however, need to analyze each problem before applying those techniques.

Example 1
68% of those polled were in favor of the Park Initiative. If 6,275 people were polled, how many were in favor of the initiative? How many were not in favor of the initiative?

Analyze: 1) The number in favor is 68% of the 6,275 people.

2) Subtract the number in favor from the total to get the number not in favor.

Solve: 1) 68% of 6,275 = 0.68 × 6275 = 4267

2) 6275 − 4267 = 2008

Example 2
According to 2001 IRS Tax Rate Schedule X (single filing status), if your taxable income after deductions is over $136,750 but not over $297,350, your tax is $36,361.00 plus 35.5% of the amount over $136,750. If your taxable income after deductions in 2001 was $175,000, how much tax did you owe?

Analyze: 1) Determine the amount of income over $136,750.

2) Find 35.5% of that amount.

3) Find total tax by adding the result of Step 2 to $36,361.00.

Solve: 1) $175,000 − $136,750 = $38,250

2) 35.5% of $38,250 = 0.355 × $38,250 = $13,578.75

3) Tax: $36,261.00
    + 13,578.75
    ─────────
    $49,939.75

COMMISSIONS

Many sales people receive all or a part of their pay as a **commission**, a percentage of their sales. The intent of a sales commission is to motivate a salesperson to sell more. To determine a person's commission we simply take a percent of the amount of sales.

Example 3
If a realtor receives a 3.5% commission on the sale of a house, how much does she earn on the sale of a $270,000 house?

Analyze: Take 3.5% of the $270,000.

Solve: 3.5% of $270,000 = 0.035 × $270,000 = $9450

PERCENT PROPORTION

In other types of percent problems, you may have to use the percent proportion.

$$\frac{A}{B} = \frac{P}{100} \quad \text{where} \quad \begin{cases} A = \text{the amount} \\ B = \text{the base} \\ P = \text{the percent} \end{cases}$$

You will find it easier, if you can translate the problem into a condensed form, **A** is **P**% of **B**. By doing that, you can readily identify the A, B, and P in the percent proportion.

Example 4

During the last three years the Hawks won 75% of their basketball games. If they won 69 games, how many games did they play?

Analyze: 1) This is a percentage problem where we are trying to determine, 69 is 75% of ?.

2) $P = 75$, $A = 69$, and B is unknown

Solve: Using the percent proportion,

$$\frac{A}{B} = \frac{P}{100}$$

$$\frac{69}{B} = \frac{75}{100}$$

$$75 \times B = 69 \times 100$$

$$\frac{75 \times B}{75} = \frac{6900}{75}$$

$$B = 92$$

PERCENT INCREASE AND DECREASE

Percents are also used as a standard way to describe changes in prices, costs, profits, and other quantities. You may read that unemployment decreased 1.6% or the profit for a company increased 17.5%. How are such percents determined? You will see from the following examples that in percent increase (decrease) problems, you use the percent proportion. In that proportion, A = amount of the increase (decrease), B = the base (starting) quantity, and P is unknown.

Example 5

If the price of a movie ticket increased from $7.00 to $7.50, what was the percent increase in the price of the ticket?

Analyze: 1) Determine the amount of the increase. ($A = \$7.50 - \$7.00 = \$.50$)

2) We want the percent increase from $7.00 so the base is $7.00. ($B = 7.00$)

3) The percent increase is unknown. Restating the problem in a condensed form, we get: $.50 is ? % of $7.00.

Solve:

$$\frac{A}{B} = \frac{P}{100}$$

$$\frac{.50}{7.00} = \frac{P}{100}$$

$$7.00 \times P = .50 \times 100$$

$$\frac{7 \times P}{7} = \frac{50}{7}$$

$$P. \approx 7.14\%$$

Example 6

A DVD player was reduced from $450 to $390. What is the percent reduction in the price of the DVD player?

Analyze: 1) Determine the amount of the decrease.

$$(A = 450 - 390 = 60)$$

2) We want the percent decrease from $450 so the base is 450.

$$(B = 450)$$

3) The percent decrease is unknown. Restating the problem in a condensed form, we get: 60 is ? % of 450.

Solve:

$$\frac{A}{B} = \frac{P}{100}$$

$$\frac{60}{450} = \frac{P}{100}$$

$$450 \times P = 60 \times 100$$

$$\frac{450 \times P}{450} = \frac{6000}{450}$$

$$P \approx 13.3\%$$

MARKUPS AND MARKDOWNS

In retail business, percents are also used as a basis for **marking up** (increasing) or **marking down** (decreasing) the price of an item. For example, a department store may mark up the cost of an item 40% before it is sold or mark down the retail price 30% during a sale.

Example 7

In order to make a profit, a bookstore may mark up the cost of books 40% to obtain the retail price for books sold in the bookstore. If a history book has a wholesale cost of $32.00, what is the retail price?

Analyze: 1) The amount of markup is found by multiplying the percent by the cost.

2) The retail price is the total of cost and the amount of the markup.

Solve: 1) 40% of $32 = 0.40 × $32 = $12.80

2) Retail price = cost + markup
   = $32.00 + $12.80
   = $44.80

Example 8

If the $44.80 history book in Example 7, goes on sale and is marked down 40% off that retail price, what is the sale price of the book?

Analyze: You might think that the sale price is $32 since we obtained the $44.80 by a 40% marking up on $32. However, as you will see that is not the case.

1. The amount of markdown is obtained by multiplying the percent by the retail price.

2. The sale price is the retail price minus the amount of the markdown.

Solve: 1. 40% of $44.80 = 0.40 × $44.80 = $17.92

2. Sale price = retail price − markdown
= $44.80 − $17.92
= $26.88

Example 9

The retail price of a shirt was $59.99. However, when the shirt was placed on the sales rack, it was marked down 25%. Then, at a Sidewalk Sale, the shirt was marked down an additional 30%.

a) What was the cost of the shirt after the final 30% markdown?

b) Is the final sale price equal to a 55% discount on the original price of the shirt?

a) *Analyze*: The amount of the markdown is obtained by multiplying the percent of the markdown (25%) by the retail price. The amount of the markdown is then subtracted from the retail price. That new price is marked down 30% to get the final cost of the shirt.

Solve: Price after the 25% markdown = Price − Markdown
= $59.99 − 0.25($59.99)
= $59.99 − $15.00
= $44.99

Price after the 30% markdown = Price − Markdown
= $44.99 − 0.30($44.99)
= $44.99 − $13.50
= $31.49

b) *Analyze*: To answer this question, compare the price after a 55% discount and the $31.49 sale price obtained in part a).

Solve: Price after the 55% markdown = Price − Markdown
= $59.99 − 0.55($59.99)
= $59.99 − $32.99
= $27.00

The sale price after a 55% discount is less than the price after successive 25% and 30% markdowns.

Example 10

If a $38.00 pair of jeans sells for $58.00, what percent markup is being used?

Analyze: This is really the same as a percent increase problem. The amount of the increase is $20 ($58 − $38) and the base price is $38. Using the percent proportion we get:

$$\frac{A}{B} = \frac{P}{100}$$

Solve:

$$\frac{20}{38} = \frac{P}{100}$$

$$38 \times P = 2000$$

$$P = 52.63\%$$

The techniques explained in these examples will enable you to solve applications involving percents. The problems that follow will give you more practice in working with percents.

Name _____   Date _____

EXERCISE 6.4 SET A

DIRECTIONS:

Write answers with money in appropriate form.
    Reduce fractions/ratios to lowest terms.
    Round decimals to nearest hundredth/cent.

1) What percentage of the Hoosier's annual budget goes towards entertainment if 15% goes for mortgage, 5% goes towards medical bills, 5% goes towards car insurance, 10% goes towards paying taxes, 15% goes towards paying utilities, 10% goes towards savings, 20% goes towards groceries, 5% goes towards clothing, and 10% goes towards house maintenance?

2) If 88% of drivers pass the written test on their first attempt, what percentage of drivers fails their first attempt?

3) There are two hundred employees in a company. Twenty percent of the employees are women. How many employees are women?

4) On Friday, 68% of moviegoers bought something to eat or drink from the concession stand. If the theatre had 12,500 moviegoers, how many of them bought something from the concession stand?

5) Charlotte sent 2000 invitations for the Charity Gala on March 15. By March 1st, 80% of the responses were received. How many responses were not received?

6) Joan has 420 milliliters of a chemical solution stored in the lab. Forty percent of the solution is Nitrogen and 12% is Phosphorus. How many milliliters of the solution are made up of other chemicals?

7) A furniture store advertises 25% off on all dining room sets. How much is the discount on a $2358.15 dining room set?

8) A novelty store is having a 30% sale on all Christmas items. Jennifer bought a $15.85 ornament, a $109.00 Singing Santa, and a $550 Christmas tree. How much did Jennifer save?

9) What is the final price of an $89.99 vase, if a $6\frac{3}{4}$% sales tax is charged?

10) How much did Leo pay for a $115 tool box and $287.56 worth of tools if the sales tax was $6\frac{1}{4}$%?

11) An appliance store is having a 35% discount on all ovens. What is the final price of a $625 oven?

12) Kirsten bought a motorcycle for $12,350. She put a 25% down payment. How much was the down payment?

13) On a $3456.50 bedroom set, Andy and Kim put a 15% down payment. How much do they still owe on the bedroom set?

14) Roger purchased a jet ski for $1879.00. He made a 25% down payment and paid the remaining amount in 9 months. How much was Roger's monthly payments?

15) Sharisse will take a 12-day tour of Italy costing $3455. She will make a 15% down payment and pay the remaining amount in monthly installments.

    a) How much does Sharisse still owe for the trip?

    b) If Sharisse will pay the remainder in 6 months, how much are the monthly payments?

16) Kathy earns $34,689 per year. She receives a 2.2% raise. How much will Kathy now earn?

17) Carrie's cable bill was $68.95 each month. The cable company announced a $5\frac{1}{2}$% increase in rates. How much will Carrie pay yearly for her cable?

18) Sixty-five percent of a community theatre group's expenses pay for the actors and actresses. The theatre group's expenses were $189,000 to put on the production of "Snow White and the Seven Dwarfs". How much of the expenses went to pay for the actors and actresses?

19) Jack had 50 pounds of frozen steaks. He barbequed 30 pounds at his 4th of July party. What percent of the frozen steaks did Jack barbeque?

20) Lillian and Frank bought a $150 platter for their parents. Frank paid $90. What percent did Lillian pay?

21) A baseball little league team lost 40 games in the last 2 years. If the team played 75 games (no ties), what percent did they win?

22) In a survey of figure skating fans, 360 people knew Michelle Kwan won her 9th U.S. National title. If that is 60% of those surveyed, how many fans were surveyed?

23) Hallie counted four hundred thirty-five 2004 calendars in stock. This represents 5% of all the calendars in stock.
   a) How many calendars are in stock?
   b) How many are not 2004 calendars?

24) In a survey of first-year students in college, 4200 students said they had "pulled an all-nighter" to study for a final exam. If that is 21% of the students, how many have them did not "pull an all-nighter"?

25) In a random inspection, Manuel found 10 pencil sharpeners to be defective. This represents 0.5% of all pencil sharpeners manufactured. How many pencil sharpeners were manufactured?

26) A produce manager at a large supermarket throws out 20 pounds of rotted vegetables per day. This represents 0.4% of the vegetables. How many pounds of vegetables are kept to be sold?

27) In one day, a company spends $35,455 on all binding of textbooks. The manager discards four textbooks because the binding was done improperly. This represents 0.8% of binding textbooks. How much is the cost of binding each textbook?

28) A pet store reduced all twenty gallon tanks by $30. If a twenty gallon tank costs $110, what is the discount rate?

29) The well-known chef of a restaurant decided to open his own business. The restaurant saw a loss of 6 customers per night once the word spread that the chef had left. If the restaurant has an average of 120 customers per night, what is the percent of decrease?

30) A hardware store increased the price of shovels by $1.80 when the first snow storm was predicted. The original price of the shovels was $8.90. What is the markup rate?

31) The price of milk has increased by $0.45 per gallon. If a gallon of milk costs $2.39, what is the percent of increase in milk prices?

32) A pottery store sells a 4-quart flower pot for $12.45. On Saturday the spring sale will begin and the flower pot will sell for $9.95. What is the percent of decrease?

33) Marco, an Italian delicatessen owner, buys salami at wholesale price for $5.50 per pound. He sells the salami at $7.25 per pound. In order to make a profit, the markup rate must be at least 30%. Did Marco make a profit?

34) To promote travel outside the United States, an airline is reducing the price of an airline ticket to Greece from $1100 to $920. What is the mark down rate?

35) In 2000, the We Tutor center had 200 customers. Because of aggressive advertising in 2001, the We Tutor center had 950 customers. What is the percent of increase in customers?

36) Daniel, at 295 pounds, and Judy, at 215 pounds, each decided to go on a diet to lose weight. After 6 months on the diet, Daniel weighs 215 pounds and Judy weighs 155 pounds. Which person had the greatest percent of weight loss?

37) In order for an appliance store to make a profit on refrigerator sales, the markup rate must be at least 30%. The appliance store buys refrigerators for $650 each at wholesale price. They will sell the refrigerators at $875 each.
   a) What is the markup rate of the refrigerators?
   b) The appliance store will have a sale and sell the refrigerators for $760 each. What is the markdown rate?
   c) Will the appliance store still make a profit during the sale on the refrigerators?

38) The Coffee Shop bought 2500 pounds of their best-selling coffee blend for $7750. To make a profit the markup must be at least 40%. The Coffee Shop sells the coffee at $5.55 per pound.
   a) What is the cost per pound of coffee for the Coffee Shop?
   b) What is the markup rate on the cost per pound of coffee?
   c) Did the Coffee Shop make a profit?

39) A mountain bike was on sale for 30% off the original price. Josephine bought the bike on sale for $250. What was the original price of the bike?

40) Due to a large endowment, Saturn College admitted 17,914 students for the fall semester. This includes a 6% increase in admittance from the previous fall semester. How many students were admitted to the College the previous fall semester?

EXERCISE 6.4 SET B

DIRECTIONS:

Write answers with money in appropriate form.
Reduce fractions/ratios to lowest terms.
Round decimals to nearest hundredth/cent.

1) A school's employees are 5% administrators, 4% guidance counselors, 3% custodians and kitchen workers, and 1% secretaries. What percent of the school's employees are teachers?

2) Of the 3500 men surveyed, 15% of them said that they would ask for a woman's phone number with other women present. How many men would ask for the woman's phone number?

3) Of the 5,500 five-year olds surveyed, 85% of them agreed with the statement, "Ice cream is the best dessert." How many five-year olds agreed with the statement?

4) Corey organized a family reunion in Myrtle Beach, South Carolina. He invited 400 people and 65% of them said they would attend. How many people will not attend the family reunion?

5) The sales tax in New Jersey is 7%. How much sales tax is charged on a computer costing $999?

6) What is the final price of a $440 pool table, if a $7 \frac{1}{2}\%$ sales tax is charged?

7) Gary bought a $345 lawn mower and a $199 snow blower. What was the final price if a 5.5% sales tax was charged?

8) A department store is having a 40% sale on all winter coats. How much do you save on a $164.45 winter coat?

9) Patrick put a 20% down payment on a $24,955 motor home. How much was the down payment?

10) An electronics store requires a 30% down payment on all items bought on credit. How much does Theodore owe on a $1450 plasma television after the down payment?

11) Waldo lost some weight and went on a shopping spree. He bought 5 pairs of jeans for $39.99 each, 8 shirts for $18 each and a belt for $15.50. There was a 40% discount on all jeans. Waldo also had to pay an $8 \frac{1}{2}\%$ sales tax. What was Waldo's total after the discount and sales tax?

12) Jessica earns $43,879 per year as a teacher. She receives a 3.5% raise. How much will Jessica now earn?

13) Marcelo's monthly rent was $1880 for an apartment in NYC. He was notified that his rent would increase by $4 \frac{3}{4}\%$. How much is Marcelo's new monthly rent?

14) Olivia negotiated a 3% raise for each of the next 3 years. Her salary was $68,500. How much will her yearly salary be in 3 years?

15) If you won 20 out of 35 poker hands, what percent did you win?

16) Eli's road trip is 1350 miles. He drove 450 miles in two days. What percent of the trip remains?

17) Bob saved $3200 last year. If that was 3.5% of his yearly income, how much was Bob's yearly income?

18) An amusement park opens for the season on the first weekend of May. Last year 245,000 people came to the amusement park during the first weekend. An all-day pass for adult tickets cost $35.00 and for children under 12 tickets cost $18.50.

   a) If 45.5% of the people who came to the amusement park were children, how many of the people were adults?

   b) How much total income did the amusement park make from the adult and children's tickets?

19) One liter of a chemical solution requires 20% water. If a lab assistant needs to make $33 \frac{1}{2}$ liters of the chemical solution, how many liters of water does the lab assistant need?

20) In a town council election Liza and David were the only candidates who ran. Liza received 880 of the votes. If that was 64% of the votes, how many people voted for David?

21) In a factory which manufactures sweaters, thirty sweaters were found not acceptable to sell to a retail store. If this represents 0.02% of the sweaters manufactured, how many sweaters will be sold to retail stores?

22) A toy manufacturer expects a $\frac{3}{4}\%$ failure rate for an electronic toy. In a random inspection, 12 electronic toys were found to be defective. How many electronic toys were not defective?

23) Yesterday a plant manager found 15 dented aluminum cans to be used for vegetables. This represents 0.003% of the cans produced. If it costs $25,000 to manufacture all the cans, what is the cost of manufacturing each can?

24) An auto parts store charges $90 for a car battery. For a promotional sale, the auto parts store takes $15 off the price of the battery. What is the discount rate?

25) As the manager of a candy store, Rosalia marked down the price of chocolate pretzels by $.60 per pound. The original price was $4.50 per pound. What is the mark down rate?

26) A florist decides to mark up the price of her Christmas wreaths by $3.50. If the wreaths cost was $14.50, what is the markup rate?

27) A bakery buys its supply of French bread at $0.90 per loaf. The bakery marks up each loaf of bread by $0.20. What is the markup rate?

28) A sports store buys tennis balls for $4.45 per canister and sells each canister for $6.35. What is the markup rate for the canister of tennis balls?

29) Nick bought an $18,500 car in April. By June, the car's value depreciated to $15,855. What is the rate of depreciation?

30) To compete with other coat factories, the wholesale price of a leather jacket to a retail store will be reduced from $230 to $180. What is the markdown rate?

31) Phillipa earns $54,600 per year. Phillipa received a promotion and she now earns $75,980. What was her percent of increase in pay?

32) Due to the restructuring of county lines, Tom's home now is considered to be part of Fisher County. His property taxes decreased from $8755 per year to $7655 per year. What is the percent of decrease in property taxes?

33) In order to make a profit on chocolate sales, the markup rate must be at least 30%. The Chocolate is Heaven store bought 300 boxes of chocolates for $8400. The store sells the chocolates for $35 per box.

   a) What is the cost per box of chocolate for the Chocolate is Heaven store?

   b) What is the markup rate on the cost of each chocolate box?

   c) Did the Chocolate is Heaven store make a profit?

34) A store charges a $6\frac{1}{2}\%$ sales tax. Jimmy paid $455.85 for a laser printer. What was the selling price of the laser printer?

35) A ski lodge's profit this year was $68,900. This is a 20% decrease from last year's profits due to little snow falls. How much did the ski lodge make in profits last year?

6.5 APPLICATIONS INVOLVING SIMPLE INTEREST

If you borrow money, you must pay back more than you borrowed. That extra amount that you pay back is called **interest**. The amount of interest you are charged is determined by taking a percent of the amount borrowed times the length of time of the loan. The amount borrowed is called the **principal** (P). The percent used is called the **rate** (R). The length of time for the loan is called the **time** (T). Interest on a loan can be determined using this formula:

These problems really pique my interest.

> **Interest = Principal × Rate × Time**
>
> **I = P × R × T**

Interest calculated using that formula is called **simple interest**, since it is computed on only the original principal during the time period. Banks, however, compute interest on previously earned interest, called compound interest, and use principles of compound interest in amortizing loans.

Example 1

How much interest is charged on a $500 loan for one year using a 12% yearly interest rate?

Analyze: Use the interest formula with:

P = $500
R = 12% = 0.12 per year
T = 1 year

Solve: I = P × R × T
= $500 × 0.12 × 1
= $60

In interest problems you must make sure that the rate and the time are expressed in the same units. If the rate (R) is a *yearly* rate, then the time (T) must be expressed in *years*. If the rate (R) is a *monthly* rate, then the time (T) must be expressed in *months*.

Example 2

What is the interest on $3200 at 1.5% per month for 2 years?

Analyze: Use the interest formula with:

P = $3200
R = 1.5% = 0.015 per month
T = 2 years = 24 months

Solve: I = P × R × T
= $3200 × 0.015 × 24
= $1152

Once you know how to calculate the interest on a loan, you can determine how much you must pay each month to pay off the loan. Keep in mind that when you pay off a loan you must pay back the principal *and* the interest.

Example 3

Determine the monthly payments necessary to pay off a $6300 loan for 3 years at an annual rate of 13%.

Analyze:
1) Determine the interest using the interest formula.
2) Determine the total to be paid back by adding the principal and the interest.
3) Determine the monthly payments by dividing the total by the number of months. (3 yr = 36 mo)

Solve:
1) I = P × R × T
   = $6300 × 0.13 × 3
   = $2457

2) Total = $6300 + $2457 = $8757
3) Payments = $8757 ÷ 36 = $243.25

Example 4

You are purchasing a new car with a total price of $9500. The dealer gives you a 10% discount and a $1500 trade-in on your old car. To pay off the balance you take out a loan with a 12.6% yearly rate for 42 months. To the nearest cent, what should your monthly payments be to pay off the loan?

Analyze:
1) Determine the balance by subtracting the 10% discount and the $1500 trade-in from the total price of $9500.
2) Determine the interest. P = $7050, R = 12.6% = 0.126 per yr., T = 42 months = 3.5 yrs.
3) Get the total to be repaid by adding principal and interest.
4. Get monthly payments by dividing the total by the number of months.

Solve:
1) Discount = 0.10 × 9500
            = $950
   Balance = 9500 − 950 − 1500 = $7050

2) I = P × R × T
   = $7050 × 0.126 × 3.5
   = $3109.05

3) Total = 7050 + 3109.05 = $10,159.05
4) Payments = $10,159.05 ÷ 42 ≈ $241.88

PROBLEM

1) Determine to the nearest cent the monthly payment needed to pay off a loan of $700 at 9.5% per year in two years.

Answer: $34.71

I = 700 × 0.095 × 2
  = $133
total = 700 + 133
      = $833
payments = 833 ÷ 24
         ≈ $34.71

Name _____   Date _____

EXERCISE 6.5

DIRECTIONS:

Write answers with money in appropriate form.
    Reduce fractions/ratios to lowest terms.
    Round decimals to nearest hundredth/cent.

1) Brett borrowed $15,000 to build a garage on his property. His bank offered him an annual interest rate of 4% to be repaid over 3 years. How much interest will Brett pay on this loan?

2) Marcie wants to expand the merchandise of her boutique on 5th Avenue in New York City to include high quality jewelry. She takes out a 4-year loan of $24,000 at an annual interest rate of 6%. How much interest will Marcie pay on this loan?

3) A law firm wishes to redecorate the three senior law partners' offices. The firm will borrow $12,500 at an annual interest rate of 13.5% to be repaid over 2 years. How much interest will the law firm pay on this loan?

4) An electronics store needs to increase its video camera inventory and purchases $225,000 worth of inventory. To finance the purchase, the electronics store takes out a $1\frac{1}{2}$ year loan at 5% annual interest rate. How much interest will be paid on the loan?

5) Roberto, a restaurant owner, wants to buy new dinnerware, flatware, glassware, and linens for the restaurant's grand reopening. He obtains an $85,000 loan to be paid in 9 months at an annual interest rate of 6.5%. How much interest will Roberto pay on this loan?

6) A fitness center borrowed $140,000 at an annual interest rate of $4\frac{1}{4}$% to buy exercise equipment. The loan will be repaid in 18 months. How much interest will be paid on this loan?

7) Damage was done to the basement of a bakery due to a flood. The bakery received a disaster loan of $155,000 to be repaid in 5 years at an annual interest rate of 3%.
    a) How much interest will be paid on the loan?
    b) What is the total amount to be repaid on the loan?

8) A clothing store company launches an advertising campaign to announce the addition of a children's department in their stores. The company borrows $45,500 at an annual interest rate of 7.2% to be repaid in 9 months.
    a) How much interest will be paid on the loan?
    b) What is the total amount to be repaid on the loan by the company?

9) Devon purchased a $445,000 home in Paramus, New Jersey. He put a down payment of $45,000 and borrowed the remaining amount at an annual interest rate of $4\frac{3}{4}$%. The loan is to be repaid over 25 years.
    a) What is the amount of the loan?
    b) How much is the interest on the loan?
    c) What is the total amount to be repaid on the loan?

10) Vanessa bought an $18,345 car. She made a 15% down payment and took out a loan for the remaining amount. Her bank offered her a 9.7% annual interest rate to be repaid over 5 years.
    a) What is the amount of the loan Vanessa must take out?
    b) How much interest is to be paid on the loan?
    c) What is the total amount to be repaid on the loan?

11) Norman purchases an Irish pub for $633,850. He obtains a 20-year loan at a $5\frac{1}{2}$% annual interest rate. Find the total amount to be repaid on the loan by Norman.

12) Danny owns a cab service and needs to borrow $70,000 to purchase 5 new cabs. The annual interest rate for the loan is $7\frac{3}{4}\%$ and will be repaid over $2\frac{1}{2}$ years. What is the total amount that will be repaid on the loan?

13) Victoria borrowed $510,000 to open her own equestrian center. She obtained a 10 year loan at a 12% annual interest rate.
    a) How much interest is to be paid on the loan?
    b) What is the total amount to be repaid on the loan?
    c) What will be Victoria's monthly payments?

14) Julia, a beauty salon owner, wants to expand her business to include a manicure and pedicure center. Julia borrows $22,400 at a 4.8% annual interest rate to be repaid over 3 years.
    a) How much interest is to be repaid on the loan?
    b) What is the total amount to be repaid on the loan?
    c) What will be Julia's monthly payments?

15) The We Produce Lamps company in Idaho expanded their business to the Northeastern states. The company borrowed $235,000 to buy more equipment for the increase in production. The loan terms are for 4.5 years at an $8\frac{1}{2}\%$ annual interest rate.
    a) What is the interest to be paid on the loan?
    b) How much is the total amount to be repaid on the loan?
    c) What will be the company's monthly payments?

16) The Relax With Us Spa has decided to do major renovations to attract celebrities. The renovations cost is $585,555. The Spa obtained a loan at an 11.5% annual interest rate for 10 years.
    a) What is the total amount to be repaid on the loan?
    b) What is the monthly payment on the loan for the Spa?

17) To ensure the drillers' safety, an oil drilling company buys new equipment. To do this, the company borrows $1,000,000 at a 13% annual interest rate for a period of 18 months. Find the company's monthly payments.

18) A bookstore will expand its business to include a café. The bookstore borrowed $158,000 at a $7\frac{1}{4}\%$ annual interest rate to be paid over 2 years. Find the bookstore's monthly payments.

19) A hobby store owner wants to stock 200 collectable model airplanes. The owner buys them from a warehouse at $65.45 each and will take out a loan to pay for them. His bank offered him 11.2% annual interest rate for a period of 9 months.
    a) What is the amount of the loan?
    b) How much is the monthly payment?

20) Lehi owns an office supply store. He ordered 50 desk top computers at $950 each and 50 laser printers at $450 each. He took out an 18-month loan at a 9.9% annual interest rate to pay for the order.
    a) How much is the loan Lehi has taken out?
    b) What is the monthly payment?

21) Helen and Kevin bought a $375,600 home in Rocks County. They made a 20% down payment and took out a loan for the remaining amount. We Lend Bank offered them a 30-year loan at a 4.6% annual interest rate. Need A Loan Bank offered them a 25-year loan at a 6% annual interest rate. Helen and Kevin have budgeted $2000 per month to repay the loan.
    a) What is the amount of the loan for the home?
    b) How much will the monthly payments be for the We Lend Bank?
    c) How much will the monthly payments be for the Need A Loan Bank?
    d) Which bank offered Helen and Kevin the better deal for their budget?

22) The Furniture Store is having a 30% off sale on all bedroom sets. Patricia and Edwin bought a $5345 bedroom set. There is a 6% sales tax. The couple will finance the bedroom set with a 9-month loan at an annual interest rate of 3.5%.
    a) What was the final cost of the bedroom set?
    b) How much will the monthly payments be for Patricia and Edwin?

23) A car dealership offered 15% off all used cars. Milena wants to buy a $9875 used car, but can afford monthly payments of no more than $300. The car dealership charges $365 for license and registrations fees. There is also a $7\frac{1}{2}\%$ sales tax.
    a) What was the final cost of the used car?
    b) Milena's bank offered her a 5-year loan at an annual interest rate of 7%. How much is the total amount Milena has to repay on the loan?
    c) Can Milena afford the monthly payments on the loan?

24) Due to lack of offers, the A.F. Corporation reduced an apartment building's selling price of $680,000 by 40%. With the reduced price the J.M. Corporation will buy the property. The Corporation will make an 18% down payment and take out a 15-year loan at a 13.5% annual interest rate on the remaining amount.
    a) What is the amount of the loan J.M Corporation will take out?
    b) How much will the monthly payments be for the Corporation?

25) You will stock 130 Fisher fishing rods which you sell for $125 each. Your cost will be $112.00 per fishing rod. To make the purchase, you take out a 9-month loan for $14,560 at an annual interest rate of 18.5%.
    a) Find the interest paid on the loan.
    b) Find the interest paid per fishing rod.
    c) Find the unit cost of each fishing rod (your cost per fishing rod + interest per fishing rod).
    d) If you sell all 130 fishing rods, what will your total profit be?

7.1 VARIABLES AND FORMULAS

CHAPTER 7

The basis of algebra is the use of letters to stand for numbers. These letters are called **variables** since the numbers that they represent can vary. The concept of using letters in the same way as numbers is the first thing that you have to accept and understand in the study of algebra. Consider the basic operations:

ADDITION

$x + y$ indicates that two numbers are being added. x and y represent the two addends.

Variable:
In order to understand "x" you must be very able.

SUBTRACTION

$p - k$ indicates that two numbers are being subtracted. p and k represent those numbers.

DIVISION

$f \div R$ or $\dfrac{f}{R}$ indicates that two numbers are being divided. f and R represent those numbers

MULTIPLICATION

$A \cdot B$ or AB or $(A)(B)$ indicates that two numbers are being multiplied. A and B represent the factors.

NOTE: We do not use the letter "x" to signify multiplication, since in algebra the "x" can be used as a variable to represent numbers. Multiplication is displayed in other ways. For example, 5 times k can be written:

1. using a dot: $5 \cdot k$
2. using parentheses: $5(k)$
3. using no sign between factors: $5k$

COMBINED OPERATIONS

$4x + 3y$ means that 4 times a number x plus 3 times a number y.

$\dfrac{P}{rt}$ means a number P is divided by the product of the numbers r and t.

Do not be confused by the letters. Just keep telling yourself that letters are used just like numbers, since they simply represent numbers.

1.) Pages 009–018, 037–042, 045–047, 052–059, 062–077, 080–087, 115–127, 129–132, 141–143, 147–152, 184–187, 199–203, 206–209, from *Math for College Students: Arithmetic with Introductions to Algebra and Geometry,* 5th Edition by Ronal Straszkow. © 1997 by Kendall/Hunt Publishing Company. Used with permission.

2.) Pages 215–218, 221–224, 227–230, 232–234, 238–240, 244–247, 251–261 from *Modern Mathematics,* 10th Edition by Ruric E. Wheeler and Ed R. Wheeler. © 1998 by Kendall/Hunt Publishing Company. Used with permission.

A fundamental problem in algebra involves finding the value of **algebraic expressions**—expressions that have numbers, variables, and operations. We can find the value of an algebraic expression, if we know the value of each variable and use the correct order of operations.

Example 1

If $x = 5$ and $y = 2$, then $4x + 3y = ?$

$$4x + 3y = 4 \cdot 5 + 3 \cdot 2 \qquad \text{Replacing each variable with its value.}$$
$$= 20 + 6 \qquad \text{Doing the multiplication.}$$
$$= 26 \qquad \text{Doing the addition.}$$

Example 2

If $A = 10$, $B = 6$, and $C = 7$, then $\frac{A}{2}(3B - C) = ?$

$$\frac{A}{2}(3B - C) = \frac{10}{2}(3 \cdot 6 - 7) \qquad \text{Replacing each variable with its value.}$$
$$= \frac{10}{2}(18 - 7) \qquad \text{Doing the multiplication in the parentheses.}$$
$$= \frac{10}{2}(11) \qquad \text{Doing the subtraction in the parentheses.}$$
$$= 55 \qquad \text{Doing the division and multiplication.}$$

FORMULAS

You have used variables previously in the text. In Chapter 5, you found simple interest on a loan using, $I = PRT$. In Chapter 6, you converted between Fahrenheit and Celsius temperatures using, $C = \frac{5}{9}(F - 32)$ and $F = \frac{9}{5}C + 32$. Those algebraic statements are called formulas. A **formula** states a known relationship between variables. In a formula, the letters used are usually letters from the quantities they represent. This makes it easier to remember the formula and facilitates substituting numbers for the correct variable. Formulas are used in many fields: biology, business, chemistry, physics, nursing, geometry, and many others. In this section, we will examine formulas from some of the other areas.

Example 3

A formula for distance traveled

If you travel at an average rate (speed) for a length of time, the distance you travel is given by the formula:

$$d = rt \quad \begin{cases} d = \text{distance} \\ r = \text{rate (speed)} \\ t = \text{time} \end{cases}$$

NOTE: The units used must agree. If the rate is in miles per hour (mph), then distance is in miles and time is in hours.

If you drive at an average speed of 45 mph for 4 hours, how far do you travel?

All you need to do is substitute the values, $r = 45$ and $t = 4$, into the formula and do the calculation.

$$d = rt$$
$$= 45 \cdot 4$$
$$= 220 \text{ miles}$$

Example 4

A formula for pediatric dosages—Clark's Rule

Clark's Rule is a method for estimating the appropriate pediatric dose of a medication from the adult dose based on the weight of the child.

$$D = \frac{aw}{150} \quad \begin{cases} D = \text{dosage for a child} \\ a = \text{adult dosage} \\ w = \text{weight of child in pounds} \end{cases}$$

If the recommended daily adult dosage of Ceftin is 600 mg, what would the daily dosage be for a child weighing 45 pounds?

To find D you need to substitute the values, $a = 600$ mg and $w = 45$, into the formula and do the calculation.

$$\begin{aligned} D &= \frac{aw}{150} \\ &= \frac{600 \cdot 45}{150} \\ &= \frac{27{,}000}{150} = 180 \text{ mg} \end{aligned}$$

Name _____   Date _____

EXERCISE 7.1 SET A

In problems 1–10, if $A = 2$, $B = 7$, and $C = 1$, evaluate each expression.

_____   1) $A + B - C$ _____   2) $B + C - A$

_____   3) $\dfrac{4B}{A}$ _____   4) $\dfrac{7A}{B}$

_____   5) $2BC - 7A$ _____   6) $14AC - 4B$

_____   7) $A(3B + 5A)$ _____   8) $A(9C + 2B)$

_____   9) $13 - 3C + \dfrac{30}{A}$ _____   10) $10 - 4A + B^2$

_____   11) **A formula for the price after sales tax**
When purchasing an item, the final price is the sum of the cost of the item and the sales tax.

$$P = C + rC \quad \begin{cases} P = \text{final price} \\ C = \text{cost of item} \\ r = \text{tax rate } (\%) \end{cases}$$

If you purchase a DVD player for $259.99 and the sales tax rate is 6.5%, what is the price after sales tax?

_____   12) Use the formula for Body Mass Index given in Example 5, to determine if a person that is 5′2″ tall and 143 pounds is considered overweight by the National Institute of Health.

_____   13) **A formula to determine skid distance**
To come to a quick stop, you apply 100% pressure to your car's brakes causing the car to skid. The distance you skid can be approximated by:

$$d = \dfrac{s^2}{30f} \quad \begin{cases} d = \text{distance in feet} \\ s = \text{speed in mph} \\ f = \text{drag factor of road} \end{cases}$$

Traveling at 50 mph, how much further do you skid on snow with a drag factor of 0.30 then on asphalt with a drag factor of 0.75?

EXERCISE 7.1 SET B

Find each answer. Follow the Order of Operations.

1) $(-8)^2 \div -4 \cdot -2 \div -1$

2) $\left(\dfrac{2}{3}\right)^2 - (-3)^4$

3) $(0.4)(0.3 - 2.3) - 0.5(12 - 8.5)$

4) $36 - 42 - 2^4 \div 0.5 + 4$

5) $\dfrac{1}{2}(5 + 12) - \dfrac{2}{3}(12 \div 5)$

6) $(0.5)^3 - 48 \div (-2)^3 - 3\dfrac{1}{2}$

Evaluate. Let $x = 4$, $y = -1$, $z = 2$, $a = \frac{1}{2}$, $b = -0.5$

7) $x - y - z$

8) $ax - by$

9) $z^2 + y(x + b)$

10) $3xy \div z$

11) $y + x \div z(a + b)$

12) $25 - ay - \dfrac{45}{y - z}$

13) $\dfrac{5xy}{2bz}$

14) $\left(\dfrac{z}{x}\right)^2 - \left(\dfrac{6y}{x}\right)$

15) The following formula converts the temperature from Celsius Degrees to Fahrenheit Degrees.

$$F = \dfrac{9}{5}C + 32 \quad C = \text{Celsius Degree} \quad F = \text{Fahrenheit Degree}$$

Roxanne will travel to Rome, Italy in June. The guide book states that the temperature at that time of year is 26° Celsius. What is the temperature in Fahrenheit Degrees?

16) The following formula gives you the total amount to be repaid to a bank when obtaining a loan.

A = P + PRT   A = Total Repaid, P = Principal, R = Interest Rate, T = Time (years)

Find the total amount to be repaid on a $30,000 loan at $2\frac{1}{2}\%$ annual interest over 3 years.

7.2 PROPERTIES OF ALGEBRA

When finding sums or products, students may switch the order of terms or group them differently so that they can easily calculate the answer.

Example: Find $5 + 21 + 15 + 9$.

Some students may add the numbers in the order that it is given.

Other students may group $5 + 15 \, (= 20)$ and $21 + 9 \, (= 30)$; and so $20 + 30 = 50$. This regrouping of numbers allows for quick mental math.

The second approach in the above example is a "legitimate" way to find the answer. That is because there are properties in algebra to support the different approaches.

COMMUTATIVE PROPERTY

Addition: $\quad a + b = b + a$

Multiplication: $a \cdot b = b \cdot a$

a, b are real numbers

The commutative property permits you to interchange the addends or factors without changing the value of the expression.

Examples:
$$10 + 5 = 5 + 10 \quad\quad\quad 10 \cdot 5 = 5 \cdot 10$$
$$15 \;\; = \;\; 15 \quad \text{and} \quad 50 \;\; = \;\; 50$$

NOTE: The commutative property does not apply to subtraction or division.

Examples:
$$10 - 5 \neq 5 - 10 \quad\quad\quad 10 \div 5 \neq 5 \div 10$$
$$5 \;\; \neq \;\; -5 \quad \text{and} \quad 2 \;\; \neq \;\; 0.5$$

ASSOCIATIVE PROPERTY

Addition: $\quad (a + b) + c = a + (b + c)$

Multiplication: $(a \cdot b) \cdot c = a \cdot (b \cdot c)$

a, b, c are real numbers

The associative property permits you to group different sets of numbers and apply the operation without changing the value of the expression.

Examples: $(10 + 5) + 8 = 10 + (5 + 8) \quad$ and $\quad (10 \cdot 5) \cdot 8 = 10 \cdot (5 \cdot 8)$

$$15 + 8 = 10 + 13 \quad\quad\quad\quad\quad 50 \cdot 8 = 10 \cdot 40$$
$$23 = 23 \quad\quad\quad\quad\quad\quad\quad 400 = 400$$

NOTE: The associative property does not apply to subtraction or division as in the commutative property.

> **IDENTITY PROPERTY**
>
> Addition: $\quad a + 0 = 0 + a = a$
>
> Multiplication: $a \cdot 1 = 1 \cdot a = a$
>
> *a is any real number*

The identity property of addition states that any number or expression added by zero is the number or expression itself. **Zero is called the additive identity.**

$$\textit{Examples:} \quad -45 + 0 = -45 \quad \text{and} \quad 0 + (x + y) = x + y$$

The identity property of multiplication states that any number or expression multiplied by one is the number or expression itself. **One is called the multiplicative identity.**

$$\textit{Examples:} \quad -45 \cdot 1 = -45 \quad \text{and} \quad 1 \cdot (x + y) = x + y$$

> **INVERSE PROPERTY**
>
> Addition: $\quad a + (-a) = (-a) + a = 0$
>
> Multiplication: $\quad a \cdot \dfrac{1}{a} = \dfrac{1}{a} \cdot a = a$
>
> *a is any real number*

The inverse property of addition states that a number or expression added to its **additive inverse or opposite** is 0.

$$\textit{Examples:} \quad 45 + (-45) = 0 \quad \text{and} \quad -xy + xy = 0$$

The inverse property of multiplication states that a number or expression multiplied by its **multiplicative inverse or reciprocal** is 1.

$$\textit{Examples:} \quad -45 \cdot -\dfrac{1}{45} = 1 \quad \text{and} \quad \dfrac{1}{(x+y)} \cdot (x+y) = 1$$

> **DISTRIBUTIVE PROPERTY**
>
> $a(b + c) = ab + ac$
>
> $a(b - c) = ab - ac$
>
> *a,b,c are real numbers*

The distributive property permits you to multiply over addition or subtraction only. It also states that you multiply to each term in the parenthesis.

$$\begin{array}{lll}
\textit{Examples:} & 5(12 - 5) & \text{and} \quad 3(x + y + 2) \\
& 5(12) - 5(5) & \quad\quad\quad 3(x) + 3(y) + 3(2) \\
& 60 - 25 = 35 & \quad\quad\quad 3x + 3y + 6
\end{array}$$

PRACTICE PROBLEMS

State the property used. Write if the property is addition or multiplication.

1) $3n + 0 = 3n$ Identity Property of Addition

2) $3y + 7 = 7 + 3y$ Commutative Property of Addition

3) $(y \cdot 5) \cdot 12 = y \cdot (5 \cdot 12)$ Associative Property of Multiplication

4) $(4 + x) + 7 = (x + 4) + 7$ Commutative Property of Addition

NOTE: Do not assume the property is associative because you see parenthesis. Take notice of what is changing on the left side to the right side. In this problem, the order of the numbers changed which is the commutative property.

5) $0 = 6abc + (-6abc)$ Inverse Property of Addition

6) $5x + 15y + 20z = 5(x + 3y + 4z)$ Distributive Property

Name _____ Date _____

EXERCISE 7.2 SET A

State the property used. Write if the property is addition or multiplication.

1) $2x + 3y + 5z = 2x + 5z + 3y$

2) $-6x + 6x = 0$

3) $(-215) \cdot 1 = -215$

4) $(4a)(5b) = (5b)(4a)$

5) $2(7m - 2) = 14m - 4$

6) $\left(\dfrac{1}{2} \cdot x\right) \cdot 2 = \dfrac{1}{2} \cdot (x \cdot 2)$

7) $\left(\dfrac{1}{3}\right)(3) = 1$

8) $(10 + 5a) + 7b = (5a + 10) + 7b$

9) $\left(\dfrac{2}{5}\right) + \left(-\dfrac{2}{5}\right) = 0$

10) $(2x - 5) \cdot 8 = 8(2x - 5)$

11) $2m + 2n - 2p = 2(m + n + p)$

12) $\left(x + \dfrac{1}{2}\right) + \dfrac{3}{8} = x + \left(\dfrac{1}{2} + \dfrac{3}{8}\right)$

EXERCISE 7.2 SET B

State the property used. Write if the property is addition or multiplication.

1) $(7 + 3x) \cdot 10 = 10(7 + 3x)$

2) $\frac{1}{4}x + \frac{1}{4}y + \frac{1}{4}z = \frac{1}{4}(x + y + z)$

3) $\left(a + \frac{-2}{3}\right) + \frac{1}{5} = a + \left(\frac{-2}{3} + \frac{1}{5}\right)$

4) $(3m + 2n + 4p) + 0 = 3m + 2n + 4p$

5) $(-6x)(1) = -6x$

6) $(-21) \cdot \left(-\frac{1}{21}\right) = 1$

7) $(3y - 4)(y + 5) = (y + 5)(3y - 4)$

8) $-3(a - 2b + 5c) = -3a + 6b - 15c$

9) $\left(\frac{1}{2} \cdot x\right) \cdot 2 = \left(x \cdot \frac{1}{2}\right) \cdot 2$

10) $\left(-\frac{8}{3}\right) + \left(\frac{8}{3}\right) = 0$

11) $(7 + 12) + 13 = (12 + 7) + 13$

12) $(-5 \cdot -4) \cdot -3 = -5 \cdot (-4 \cdot -3)$

7.3 SIMPLIFY ALGEBRAIC EXPRESSIONS

In past sections we have simplified expressions with real numbers using order of operations. In this section, we would like to begin simplifying expressions with variables.

A ***variable*** is a symbol to represent a real number. The symbol is usually a letter from the English alphabet or Greek alphabet.

    A ***term*** is a number, a variable, or the product of a number and variables.

        Examples: 5, x, $-2x$, $5x^3y^2$ are all referred to as terms.

A ***coefficient*** of a term is the real number multiplying the variable.

        Examples: The coefficient for the term 2x is 2.

                           The coefficient for the term $-\frac{2}{3}x^2y^3$ is $-\frac{2}{3}$.

                           The coefficient for the term x is an understood positive one.

Note: The sign is included when identifying the coefficient.

A ***constant*** refers to a real number (no variable) in an expression.

        Examples: In the expression, $5x + 2y - 6z + 8$, 8 is called the constant.

An ***algebraic expression*** is the sum of difference of terms.

        Examples: $3x - 5y + 2z - 10$ is an expression.

                           $4x^2y + 2x^3$ is an expression.

Like terms are terms in which the variable part is the same. The coefficients may differ.

        Examples: $3x, -5x, 1.3x$, and x are like terms.

                           $-4a^2b^3$, $\frac{1}{2}a^2b^3$, and $25a^2b^3$ are like terms.

                           $5mn^2$ and $-4m^2n$ are *not* like terms.

The above vocabulary is important to know as these are the terms used when discussing algebraic expressions.

The instruction of "collect like terms" or "combine like terms" or "simplify" tells you to add or subtract the coefficients of like terms. When collecting like terms, you are using the commutative property and associative property, whether you write the step down or mentally do the step.

Recall the rules for addition and subtraction with signed numbers given in previous sections.

PRACTICE PROBLEMS

Collect like terms.

    1) $-2x + 5x - 6x$ given expression; The terms are all like terms.

          $3x - 6x$ Therefore, add/subtract the coefficients.

            $-3x$

    2) $7x + 3y - 8x + 4y + 10x - 3y$ given expression

          $(7x - 8x + 10x) + (3y + 4y - 3y)$ group like terms

             $9x + 4y$ add/subtract the like terms

NOTE: In problem 2, the answer can also be written as $4y + 9x$. This is because of the commutative property of addition. Therefore, when using an answer key to check your answers and the terms are in different order, you know that the answers are equal.

3) $\frac{1}{2}x^2 - 9x + 3x - x^2 - 4.5x + \frac{1}{3}x^2$ given expression

$\left(\frac{1}{2}x^2 - x^2 + \frac{1}{3}x^2\right) + (-9x + 3x - 4.5x)$ group like terms

$-\frac{1}{6}x^2 - 10.5x$ add/subtract the like terms

The above three problems are straightforward; add or subtract the like terms. However, once we begin to place grouping symbols in algebraic expressions, the problems become more challenging, but not impossible.

The instruction of "simplify the expression" tells you to eliminate grouping symbols and collect like terms while following order of operations.

GUIDE TO SIMPLIFY EXPRESSIONS

- Simplify grouping symbols using the distributive property. If the expression has more than one grouping symbol, then simplify from inner grouping symbols to outer grouping symbols.
- Combine like terms using addition and subtraction.

PRACTICE PROBLEMS

Simplify the algebraic expressions.

1) $3(x + 5) + 2(3x - 6)$ given expression
$3x + 15 + 6x - 12$ use the distributive property
$(3x + 6x) + (15 - 12)$ group like terms
$9x + 3$ add/subtract the like terms

2) $-(3a - 2b) + 3a - 5b$ given expression
$-3a + 2b + 3a - 5b$ use the distributive property; an understood negative one is the multiplier
$(-3a + 3a) + (2b - 5b)$ group like terms
$0 - 3b$ add/subtract the like terms
$3b$ final answer

3) $2[x - (3 - x)]$
$2[x - 3 + x]$ use the distributive property within the brackets
$2[2x - 3]$ combine like terms within the brackets
$4x - 6$ use the distributive property

NOTE: There may be more than one way to reach the same solution as long as you are correctly following operations. Problem 3 will be simplified by a different approach below.

$2[x - (3 - x)]$ given expression
$2[x - 3 + x]$ use the distributive property within the brackets
$2x - 6 + 2x$ use the distributive property again
$4x - 6$ add/subtract like terms

4) $9(x - 5) - 3(8 + 4x) + 3[x - (4 + x)]$ given expression
$9(x - 5) - 3(8 + 4x) + 3[x - 4 - x]$ distribute within brackets
$9x - 45 - 24 - 12x + 3x - 12 - 3x$ distribute to the parenthesis
$-3x - 81$ add/subtract like terms

5) $15 - 11\{2 - 5[3a + 4b + 2(4a - 3b + 6)]\}$ given expression
$15 - 11\{2 - 5[3a + 4b + 8a - 6b + 12]\}$ distribute within brackets
$15 - 11\{2 - 5[11a - 2b + 12]\}$ add/subtract like terms within brackets
$15 - 11\{2 - 55a + 10b - 60\}$ distribute within braces
$15 - 11\{-55a + 10b - 58\}$ add/subtract like terms within braces
$15 + 605a - 110b + 638$ distribute to the brace
$653 + 605a - 110b$ add/subtract like terms

Name _____   Date _____

EXERCISE 7.3 SET A

Simplify each algebraic expression.

1) $7x - 4x + 3x + 10x - 15x$

2) $2a + 12b - 8b + 9a - 12a + 3b$

3) $9x^2 - 4y^2 + 15xy - y^2 - 7xy + 20x^2$

4) $45m + 2(3 + 18m)$

5) $6(2x + y) - 16x + 4y$

6) $27 - 8(a - 5)$

7) $8(3a - 4b) + 9(b - a)$

8) $\dfrac{1}{3}(6x - 15y + 36z) + 2(x + y - z)$

9) $50 + 10(2x - 5) - (17x - 10)$

10) $16m - 4(m + 5n - 2mn) + 3mn + 2(3n - 7m)$

11) $30x + 2[5 + 7(x - 3)]$

12) $21 - 3[4 - (y + 5) + 2(3y + 1)]$

13) $36x + 2[10x + 2(11y - 14) - 7(6y - 4)]$

14) $\dfrac{2}{3}\left[5 - \dfrac{1}{3}(4x + 6) + \dfrac{1}{2}(3x - 12)\right]$

15) $45 + 2\{3a - [5(4a - 3) + 25a + 6(5a - 4)]\}$

16) $24y - 4\{2[2y - (3 - 5y) + 6] - 8(y + 1)\}$

17) $10 - [3 + 6(x + 8) - (41 + 6x)]$

18) $2[4n - 3(2n + 2) - 1] - [n + 6(2 - n)]$

19) $6 - 8\{4b - 11(3b - 2a) - 5a + 11(2b - 3a)\}$

20) $3x + 5\{7x + 2[9 - (2x - 9) + 3(6 - x)] - x\}$

EXERCISE 7.3 SET B

Simplify each algebraic expression.

1) $8y + 3y - 4y + 10y - 15y - 11y$

2) $12m - 4n + 3n - 6 - 7m + 2 - 5n$

3) $6x^2 - 3y^2 + 16xy - y^2 - 9xy + 13x^2$

4) $21a + 5(2 + 11a)$

5) $7(3x - 2y) + 18x + 4y$

6) $32 - 6(b - 4)$

7) $3(4x - 2y) - 8(y - x)$

8) $\dfrac{1}{4}(16x - 4y + 24z) + \dfrac{2}{3}\left(3x + 12y - \dfrac{3}{2}z\right)$

9) $65 + 15(3x - 4) - (27x - 20)$

10) $14a - 5(a + 4b - 3ab) + 2ab + 3(3a - 7b)$

11) $25y - 2[8 + 6(y - 4)]$

12) $31 - 4[5 - (x + 7) + 2(8x + 1)]$

13) $42x + 3[10x + 5(11x - 10) - 6(7x - 4)]$

14) $\dfrac{3}{4}\left[8 - \dfrac{1}{4}(4x + 8) + \dfrac{1}{2}(5x - 14)\right]$

15) $34 + 3\{2a - [6(4a - 1) - 35a - 4(3a - 4)]\}$

16) $12b + 2\{2[3b - (2 - 6b) + 5] - 8(b + 4)\}$

17) $7 + [6 + 3(x + 9) - (40 + 5x)] + 2x$

18) $3[m - 3(8m + 2) - 1] - [5m + 4(3 - m)]$

19) $9 - 10\{2x - 12(3x - 4y) - 5x + 10(2x - 3y)\}$

20) $5y - \{8y + 2[7 - (9y + 3) + 6(3 - y)] - y\}$

7.4 SOLVING EQUATIONS USING THE ADDITION PROPERTY AND MULTIPLICATION PROPERTY

Now that we have covered the operations with signed numbers, we can return to the use of variables in algebra. In Section 7.1, you learned what a variable is and how to find the value of an algebraic expression. The next and probably the most important process in elementary algebra is learning how to solve algebraic equations.

An **equation** is a statement that two quantities are equal. The value of the terms on the left of the equals sign (the left side) is equal to the value of the terms on the right of the equals sign (the right side). The following are examples of equations:

$$y - 5 = 2$$
$$X + 7 = -3$$
$$21 = 4P + 1$$
$$-\frac{1}{2}R = 8$$

In each of the above equations, a variable is used. We know that a variable is simply a letter that represents a number. In an equation, the objective is to determine what number the variable represents.

Let's try to determine what number the variable represents in each of the above equations.

1) $y - 5 = 2$ What value for y will make $y - 5$ equal 2?
   $y = 7$ since $7 - 5$ gives an answer of 2.

2) $X + 7 = -3$ What value for X will make $X + 7$ equal -3?
   $X = -10$ since $-10 + 7$ gives an answer of -3.

3) $21 = 4p + 1$ What value for p will make $4p + 1$ equal 21?
   $5 = p$ since $4 \cdot 5 + 1$ gives an answer of 21.

4) $-\frac{1}{2}R = 8$ What value for R will make $-\frac{1}{2}$ times R equal 8?
   $R = -16$ since $-\frac{1}{2}$ times -16 gives an answer of 8.

What we have done in the previous examples is solve for the variable in each equation. We determined the value for the variable that made the left side equal to the right side of each equation, without using any specific method to find the answers. We looked at each equation and mentally figured out the answers. We solved them by inspection. See if you can determine the solutions to the following equations in the same manner.

Property:
A sophisticated English drink.

The inspection method for solving equations as seen in the previous section works as long as you can mentally determine the correct value for the variable in the equation. As the problems get more involved, it will be very difficult to solve the equation just by mental inspection. In this and in the next two sections, we will develop a method for solving equations that have one variable.

An equation is like a balance. If you add the same amount to both sides of the balance, it will still be in balance.

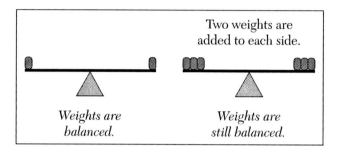

What this means is that in an equation you can add the same amount to both sides of the equation and the results will be equal. That fact is known as:

THE ADDITION PROPERTY OF EQUALITY

For any numbers A, B, and C, if $A = B$, then $A + C = B + C$.

You can use this property to help solve equations.

Example 1
Solve for P.

$$P - 7 = 19$$
$$+ 7 \quad +7$$
$$\overline{P + 0 = 26}$$
$$P = 26$$

If you add 7 to both sides of the equation, you get P on the left hand side and the answer for the variable (26) on the right.

(NOTE: $-7 + 7 = 0$ and $P + 0 = P$ on the left side of the equation.)

Example 2
Solve for X.

$$X + 5 = -3$$
$$- 5 \quad -5$$
$$\overline{X \quad = -8}$$

If you add -5 to both sides, you will have X on the left side and the answer (-8) on the right.

The method, then, for solving equations is to go through a series of steps so that the variable is alone on one side of the equals sign and a number is on the other side. That number will be the solution for the variable in the equation. Adding the same quantity to both sides of an equation is sometimes all you need to do to find the number which the variable represents.

Example 3
Solve for g.

$$-5.3 + g = 12.8$$
$$+5.3 \qquad +5.3$$
$$\overline{\qquad g = 18.1}$$

Add 5.3 to both sides, get g on the left, and the answer (18.1) on the right.

Example 4

Solve for R.

$$-37 = R + 16$$
$$\underline{-16 \quad\quad -16}$$
$$-53 = R$$

Add -16 to both sides, get R on the right, and the answer (-53) on the left.

PROBLEMS

1) Solve for B.

$$B + \frac{1}{2} = \frac{1}{4}$$

Answer: $B = -\frac{1}{4}$

$$B + \frac{1}{2} = \frac{1}{4}$$
$$\underline{-\frac{1}{2} \quad\quad -\frac{1}{2}}$$
$$B = -\frac{1}{4}$$

2) Solve for k.

$$187 = 79 + k$$

Answer: $K = 108$

$$187 = 79 + k$$
$$\underline{-79 = -79}$$
$$108 = k$$

3) Solve for T.

$$-67 + T = -19$$

Answer: $T = 48$

$$-67 + T = -19$$
$$\underline{+67 \quad\quad +67}$$
$$T = 48$$

NOTE: You can always check your answer by replacing the variable with the answer you obtained to see if it, in fact, makes the left side equal to the right side of the equation.

Not all equations can be solved by adding a number to both sides of the equals sign. The Addition Property only works in an equation that has a number added to or subtracted from the variable. There are other equations where you have to multiply or divide both sides of the equation by a number in order to obtain the variable on one side and the answer on the other side.

If you multiply or divide both sides of an equation by the same number, both sides are still equal. These properties can be stated as follows:

THE MULTIPLICATION PROPERTY OF EQUALITY

For any numbers A, B, and C, if $A = B$, then $A \cdot C = B \cdot C$.

THE DIVISION PROPERTY OF EQUALITY

For any numbers A, B, and C, if $A = B$, then $\frac{A}{C} = \frac{B}{C}$

NOTE: $\frac{A}{C}$ is another way to write $A \div C$. C can not be equal to zero since we can not divide by zero

Here is how those properties are used to solve equations:

Example 1
Solve for y.

$$6y = -30$$

Divide both sides by 6, leaving y on the left side, and the answer (-5) on the right.

$$\frac{6y}{6} = \frac{-30}{6}$$

(Note: $\frac{6y}{6} = 1y = y$)

$$y = -5$$

Example 2
Solve for x.

$$\frac{1}{3}x = 9$$

Multiply both sides by 3, leaving the x on the left, and the answer (27) on the right.

$$3 \cdot \frac{1}{3}x = 9 \cdot 3$$

(Note: $3 \cdot \frac{1}{3}x = 1x = x$)

$$x = 27$$

Example 3
Solve for V.

$$-24 = -4V$$

Divide both sides by -4, leaving V on the right, and the answer (6) on the left.

$$\frac{-24}{-4} = \frac{-4V}{-4}$$

$$6 = V$$

Example 4
Solve for m.

$$\frac{m}{7} = 8$$

Multiply both sides by 7, leaving m on the left, and the answer (56) on the right.

$$7 \cdot \frac{m}{7} = 8 \cdot 7$$

$$m = 56$$

The Multiplication and Division Properties enable you to solve equations where the variable is divided or multiplied by a number. You must remember that you want to get the variable alone on one side of the equation. If you have a number *times* the variable, you must *divide* both sides by the number. If you have the variable *divided* by a number, you must *multiply* both sides by that divisor.

PROBLEMS

1) Solve for k.

   $-8k = 19$

   Answer: $k = -2\frac{3}{8}$

   $$\frac{-8k}{-8} = \frac{19}{-8}$$
   $$k = -2\frac{3}{8}$$

2) Solve for d.

   $\dfrac{d}{6} = -3$

   Answer: $d = -18$

   $$6 \cdot \frac{d}{6} = -3 \cdot 6$$
   $$d = -18$$

3) Solve for R.

   $-9 = -2R$

   Answer: $R = 4\frac{1}{2}$

   $$\frac{-9}{-2} = \frac{-2R}{-2}$$
   $$4\frac{1}{2} = R$$

4) Solve for H.

   $\dfrac{1}{5}H = 17$

   Answer: $H = 85$

   $$5 \cdot \frac{1}{5}H = 17 \cdot 5$$
   $$H = 85$$

Name _____   Date _____

EXERCISE 7.4 SET A

Determine if the given value for the variable is the solution to the equation.

1) $3x + 5x - 2x = 12$ if $x = 2$

2) $10x - 20 = 30$ if $x = 10$

3) $\frac{2}{3}x - 6 = 12$ if $x = -3$

4) $5(x + 4) = 20$ if $x = 0$

5) $6 - x + 2 = 3x - 4$ if $x = -3$

6) $12 - 2(5 + 3x) = 6(x - 4) + 10$ if $x = 1\frac{2}{11}$

Solve each equation using the Addition Property of Equality or the Multiplication/Division Property of Equality.

7) $x + 21 = 31$

8) $x - 12 = 44$

9) $x - 55 = 45$

10) $x + 15 = -21$

11) $28 + x = 14$

12) $-13 + x = -8$

13) $90 = x - 20$

14) $-32 = x + 22$

15) $180 = 150 + x$

16) $-75 = 25 + x$

17) $x - \frac{2}{5} = \frac{1}{5}$

18) $\frac{3}{4} + x = 3$

19) $-\frac{5}{3} = x - 4$

20) $5\frac{1}{4} = -2\frac{3}{5} + x$

21) $36x = 72$

22) $-12x = 108$

23) $-55 = -5x$

24) $424 = -4x$

25) $\frac{2}{3}x = -8$

26) $-5x = -\frac{1}{5}$

27) $\frac{21}{25} = -\frac{3}{10}x$

28) $\frac{1}{3} = 4x$

29) $2\frac{3}{4}x = -1\frac{1}{2}$

30) $-\frac{32}{18} = -\frac{16}{9}x$

EXERCISE 7.4 SET B

Determine if the given value for the variable is the solution to the equation.

1) $8y + 4y - 6y = 24$ if $y = 4$

2) $20y - 40 = 60$ if $y = 6$

3) $-\frac{1}{4}y - 8 = 12$ if $y = -4$

4) $7(4 - y) = 28$ if $y = 0$

5) $8 - y + 2 = 4y - 6$ if $y = -1$

6) $10 - 6(3 + 3y) = 2(y - 4) + 1$ if $y = 1\frac{2}{11}$

Solve each equation using the Addition Property of Equality or the Multiplication/Division Property of Equality.

7) $y - 33 = 23$

8) $y + 16 = 45$

9) $y - 21 = 14$

10) $y - 11 = -21$

11) $30 + y = -9$

12) $-15 + y = -10$

13) $88 = y - 59$

14) $-42 = y - 18$

15) $-100 = 250 + y$

16) $64 = 37 + y$

17) $y - \frac{1}{8} = -\frac{3}{8}$

18) $\frac{1}{5} + y = -4$

19) $-\frac{7}{2} = y - 10$

20) $-3\frac{2}{3} = -8\frac{4}{9} + y$

21) $72y = 36$

22) $-11y = 121$

23) $-39 = -3y$

24) $235 = -5y$

25) $\frac{3}{2}y = -6$

26) $-15y = -\frac{1}{15}$

27) $\frac{36}{27} = -\frac{9}{10}y$

28) $\frac{4}{3} = 2y$

29) $-5\frac{1}{4}y = -2\frac{1}{2}$

30) $-\frac{9}{18} = -\frac{54}{36}y$

7.5 MORE ON SOLVING EQUATIONS

In the last two sections, we covered the basic properties that are needed to solve equations with one variable. In those sections, we used just one of the properties to obtain the answer for each equation. Some equations, however, require the use of more than one property to get its solution. The most efficient order to apply those properties is:

First: If a number is added to or subtracted from a variable term, use the Addition Property to get that variable term alone on one side of the equation.
Second: If a variable is divided or multiplied by a number, use the Multiplication or Division Property to get the variable equal to its answer.

THE GOLDEN RULE FOR SOLVING EQUATIONS

"Do unto one side as you do unto the other."

Example 1
Solve for x.

$$5x - 4 = 31$$

$$\begin{array}{r} 5x - 4 = 31 \\ +4 \quad +4 \end{array}$$ Add 4 to both sides, leaving the variable term ($5x$) on the left side.

$$5x = 35$$

$$\frac{5x}{5} = \frac{35}{5}$$ Divide both sides by 5, leaving x on the left, and the answer (7) on the right.

$$x = 7$$

Example 2
Solve for n.

$$6 - 3n = 18$$

$$\begin{array}{r} 6 - 3n = 18 \\ -6 \quad\quad -6 \end{array}$$ Add -6 to both sides, to get the variable term ($-3n$) on one side.

$$-3n = 12$$ Divide both sides by -3, leaving n on the left, and the answer (-4) on the right.

$$\frac{-3n}{-3} = \frac{12}{-3}$$

$$n = -4$$

Example 3
Solve for x.

$$\frac{2x}{3} = 16$$

$$(3)\frac{2x}{3} = 16(3)$$ Multiply both sides by 3, leaving $2x$ on the left and 48 on the right.

$$\frac{2x}{2} = \frac{48}{2}$$ Divide both sides by 2, leaving x on the left, and the answer (24) on the right.

$$x = 24$$

Example 4
Solve for S.

$$\frac{S}{4} + 7 = 10$$

$$\frac{S}{4} + 7 = 10$$
$$\underline{-7 -7}$$ Add −7 to both sides, leaving the variable term (S/4) on the left side.
$$\frac{S}{4} = 3$$

$$4 \cdot \frac{S}{4} = 3 \cdot 4$$ Multiply both sides by 4, leaving S on the left, and the answer (12) on the right.

$$S = 12$$

Example 5
Solve for n.

$$-4 = \frac{n}{2} - 7$$

$$-4 = \frac{n}{2} - 7$$
$$\underline{+7 +7}$$ Add 7 to both sides, to get the variable term (n/2) on one side.
$$3 = \frac{n}{2}$$

$$2 \cdot 3 = \frac{n}{2} \cdot 2$$ Multiply both sides by 2, leaving n on the right, and the answer (6) on the left.

$$6 = n$$

PROBLEMS

1) Solve for z.   *Answer:* $z = 7\frac{1}{2}$

   $2z - 2 = 13$

   $$2z - 2 = 13$$
   $$\underline{+2 +2}$$
   $$\frac{2z}{2} = \frac{15}{2}$$
   $$z = 7\frac{1}{2}$$

2) Solve for N.   *Answer:* $N = 18$

   $-1 = \dfrac{N}{-3} + 5$

   $$-1 = \frac{N}{-3} + 5$$
   $$\underline{-5 -5}$$
   $$-6 = \frac{N}{-3}$$
   $$(-3)(-6) = \frac{N}{-3}(-3)$$
   $$18 = N$$

The ability to simplify algebraic expressions will enable you to find solutions to a wider range of equations. In an equation that contains expressions that look complicated, we can combine like terms, simplify expressions, and then solve the resulting equation.

Example 1

Solve for x: $2(3x + 5) - 4x = 16$

$$2(3x + 5) - 4x = 16$$
$$6x + 10 - 4x = 16 \quad \text{by the distributive property.}$$
$$6x - 4x + 10 = 16 \quad \text{by the commutative property.}$$
$$2x + 10 = 16 \quad \text{by combining like terms.}$$
$$\underline{-10 \; -10}$$
$$2x = 6 \quad \text{by subtracting 10 from both sides.}$$
$$x = 3 \quad \text{by dividing both sides by 2.}$$

Since we know how to add and subtract like terms, this technique can be used to solve equations that have variable terms on both sides of the equal sign.

Example 2

Solve for x: $x + 9 = -2x + 21$

$$x + 9 = -2x + 21$$
$$\underline{+2x \qquad\quad +2x} \quad \text{Put the variables on one side by adding } 2x \text{ to both sides.}$$
$$3x + 9 = 0 + 21$$
$$3x + 9 = 21$$
$$\underline{\quad -9 \; -9} \quad \text{Put the numbers on the other side by subtracting 9 from both sides.}$$
$$3x = 12$$
$$x = 4 \quad \text{Obtain answer by dividing both sides by 3.}$$

Combining the techniques used in the two previous examples we get the following strategy for solving equations with one variable.

SOLVING EQUATIONS

1) Simplify each side of the equation.
2) Put the variable terms on one side of the equation by using the addition or subtraction property.
3) Put the numbers on the other side of the equation by using the addition or subtraction property.
4) Obtain the answer for the variable by using the multiplication or division property.

While using this strategy, it is important to do one step at a time and to make sure your computations with signed numbers are correct. Let's show you how this strategy works by solving a few more equations.

Example 3

Solve for x: $5x - 3x = 9 + x - 4$

$$5x - 3x = 9 + x - 4$$
$$2x = 9 - 4 + x \quad \text{Simplify the left and right side.}$$
$$2x = 5 + x$$
$$\underline{-x \qquad\quad -x} \quad \text{Put the variables on one side by subtracting } x \text{ from both sides.}$$
$$x = 5$$

Example 4

Solve for x: $4(x - 3) + 7 = 6x + 9$

$$
\begin{array}{rl}
4(x - 3) + 7 = & 6x + 9 \\
4x - 12 + 7 = & 6x + 9 \\
4x - 5 = & 6x + 9 \\
-6x \quad\quad & -6x \\
\hline
-2x - 5 = & \quad 9 \\
+5 & +5 \\
\hline
-2x & \quad 14 \\
x = & -7
\end{array}
$$

Simplify the left side.

Put the variables on one side by subtracting $6x$ from both sides.

Put the numbers on the other side by adding 5 to both sides.

Obtain answer by dividing by -2.

The value we have found for the variable in each equation will make the right side of the equation equal the left side of the equation. There is little chance that we could have obtained those answers by guessing. The strategy described in this section gives an efficient means of arriving at the solution to an equation.

PROBLEMS

1) Solve for x: $2(3x - 5) = -10$ *Answer:* $x = 0$

$$
\begin{array}{r}
2(3x - 5) = -10 \\
6x - 10 = -10 \\
+10 \quad +10 \\
\hline
6x = 0 \\
x = 0
\end{array}
$$

2) Solve for x: $5x - 7 = 4x - 8$ *Answer:* $x = -1$

$$
\begin{array}{r}
5x - 7 = 4x - 8 \\
-4x \quad\quad -4x \\
\hline
x - 7 = -8 \\
+7 \quad +7 \\
\hline
x = -1
\end{array}
$$

3) Solve for x: $6 + 2x + 3x = 16 - (2 - x)$ *Answer:* $x = 2$

$$
\begin{array}{r}
6 + 2x + 3x = 16 - 2 + x \\
6 + 5x = 14 + x \\
-x \quad\quad -x \\
\hline
6 + 4x = 14 \\
-6 \quad\quad -6 \\
\hline
4x = 8 \\
x = 2
\end{array}
$$

Name _____   Date _____

EXERCISE 7.5 SET A

Solve each equation. Check your solutions.

1) $5x + 3 = 13$

2) $10x - 15 = 25$

3) $-11 + 2x = 33$

4) $4.5 - x = 9.5$

5) $43 = 8x - 13$

6) $3 = \frac{1}{2}x - 5$

7) $\frac{2}{3}x - \frac{1}{2} = \frac{5}{6}$

8) $3\frac{1}{4} = 6 + \frac{3}{4}x$

9) $5x + 7x = -108$

10) $\frac{1}{2} = \frac{5}{8}x + \frac{9}{8}x$

11) $-11x - 19x = 150$

12) $20 = 8.6x - 14.6x$

13) $3(x + 8) = 12$

14) $-2(3x - 1) = 44$

15) $-15 = \frac{2}{3}(3x - 12)$

16) $4x + 9 = 3x - 5$

17) $7 - 6x = 9x + 22$

18) $10x - 30 = 35 + 20x$

19) $\frac{4}{5}x - 12 = \frac{1}{5}x - 10$

20) $3x - \frac{2}{3} = \frac{1}{2} + 5x$

21) $7x - 23.6 = 3x + 4.4$

22) $2(3x - 5) = 3(4x + 6)$

23) $-(11 - x) = 5(x - 4)$

24) $\frac{1}{4}(16x - 4) = \frac{1}{3}(3x - 3)$

25) $0.5(8x + 12) = 0.3(10 + 10x)$

26) $7(x + 4) = 2(x + 14)$

27) $\frac{2}{5}(x - \frac{5}{8}) = \frac{3}{4}(2x - 1)$

28) $3x + 5x - 4 = 12 + 7x$

29) $15 - 2x - 25 = 3x + 4 - 4x$

30) $\frac{1}{2}x + 8 - 3 = \frac{4}{5} - 2x + \frac{1}{2}x$

31) $2(x + 3) - 4x = 16$

32) $3 - x = 4(3x - 4) + 12$

33) $5(3 - x) - 2x = 4(x + 1) - 11$

34) $\frac{2}{5}(5x - 2) + \frac{3}{5} = \frac{1}{3}(9x - 15)$

35) $-7x + 8(4x + 5) = 5x$

36) $2x + 11 - 5x = 33 - (x - 11)$

37) $0.1(2 - 4x) - 0.3 = 0.5(10x - 20)$

38) $13x - 2(7x + 9) + 12 = 19 - 4(4x - 5)$

39) $8x - 4[3(x + 1) - (2 + 3x)] = 84$

40) $5(x + 1) = 8x + 3[12 - 2(x - 1) + 3(x + 1)]$

EXERCISE 7.5 SET B

Solve each equation. Check your solutions.

1) $6y + 1 = 13$

2) $15y - 10 = 35$

3) $-12 + 8y = 60$

4) $5.7 - y = 8.7$

5) $62 = 9y - 10$

6) $-4 = \frac{1}{3}y - 4$

7) $\frac{1}{2}y - \frac{1}{3} = -\frac{5}{6}$

8) $-2\frac{2}{5} = 8 + \frac{6}{5}y$

9) $9y - 17y = -144$

10) $\frac{1}{2} = \frac{7}{9}y + \frac{1}{9}y$

11) $-21y - 29y = 150$

12) $20 = 6.1y - 10.1y$

13) $4(y - 2) = 16$

14) $-3(2y - 1) = 50$

15) $-9 = \frac{1}{3}(6y - 12)$

16) $7y + 3 = 4y - 12$

17) $5 - 9y = 3y + 29$

18) $10y - 45 = 35 + 30y$

19) $\frac{1}{4}y - 10 = \frac{1}{3}y - 9$

20) $4y - \frac{1}{5} = \frac{1}{2} + 3y$

21) $5y - 11.3 = 2y + 4.3$

22) $3(2y - 6) = 4(3y + 5)$

23) $-(14 - y) = 4(y - 4)$

24) $\frac{1}{3}(12y - 9) = \frac{1}{4}(4y - 12)$

25) $0.2(5y + 10) = 0.5(20 - 50y)$

26) $2(y + 9) = 7(y + 14)$

27) $\frac{3}{4}\left(y - \frac{12}{15}\right) = \frac{1}{2}(2y - 1)$

28) $5y + 4y - 7 = 13 + 7y$

29) $23 - 3y - 27 = 4y + 4 - 6y$

30) $\frac{2}{3}y + 3 - 8 = \frac{1}{6} - 4y + \frac{1}{2}y$

31) $4(y + 3) - 5y = 18$

32) $2 - y = 5(4y - 5) - 12y$

33) $3(4 - y) - 2y = 6(y + 2) - 11$

34) $\frac{3}{5}(10y - 2) + \frac{3}{5} = \frac{1}{5}(15y - 10)$

35) $-8y + 7(3y + 4) = 9y$

36) $4y - 12 - 5y = 55 + (y - 11)$

37) $0.3(4 - 2y) - 0.2 = 0.5(2y - 3)$

38) $14y - 3(4y + 9) + 23 = 21 - 5(y - 2)$

39) $12y - 5[4(y + 2) - (1 + 3y)] = 49$

40) $6(y + 1) = 5y + 4[1 - 3(y - 4) + 2(y + 1)]$

7.6 MORE WORDS TO ALGEBRA AND SOLVE

In an earlier section, you were given a list of key words for operations and the algebraic translation. Below is an abbreviated version to refresh your memory.

Computation	Key Word(s)		Computation	Key Word(s)
Addition	plus add sum added to more than increased by total		Multiplication	times multiplied by product twice (times 2) doubled (times 2) tripled (times 3) fractional part of
Subtraction	minus subtract from difference fewer than less than decrease by		Divide	divided by quotient
	Equality	is equal to is result is same result as		

In this section, you will translate the verbal expression into an equation and solve for the unknown. This combines two skills you have learned, words to algebra and solving an equation.

PRACTICE PROBLEMS

Write each sentence as an algebraic equation. Then solve for the unknown number.

1) The sum of twelve and a number is 20. Find the number. $12 + x = 20; x = 8$

2) The product of two-thirds and a number is 6. Find the number. $\frac{2}{3} \cdot x = 6; x = 9$

3) Two times the difference of a number and seven is 18. Find the number. $2(x - 7) = 18; x = 16$

4) Five times a number added to ten is equal to four times the number. Find the number
$5x + 10 = 4x; x = -10$

Name _____   Date _____

EXERCISE 7.6 SET A

Write each sentence as an algebraic equation. Then solve for the unknown number.

1) The sum of six and a number is twenty-one.

2) The difference of eight and a number is thirty.

3) The product of a number and three-fourths is nine.

4) The quotient of a number and fifteen is three.

5) Five less than twice a number is thirteen.

6) The product of eight and a number plus eleven is -77.

7) The quotient of three less than a number and four is one.

8) One more than three times a number is -23.

9) Twice the sum of a number and six is twelve.

10) Twenty subtracted from one-half a number is ten.

11) Thirteen plus a number is three less than twice the number.

12) The sum of twenty and twice a number gives the same result as one-half the number.

13) The sum of ten times a number, six times the number, and four times the number is -60.

14) The quotient of a number and -4, increased by twelve, is twenty-one.

15) Three-fourths the sum of a number and eight is thirty-three.

16) Twice the difference of a number and three is the same as three times the number plus nine.

17) Forty-five less than three times a number is equal to one-half the number.

18) Ten increased by four times a number is the same result as five more than the number.

19) Six times the sum of four and two-eighths of a number is three.

20) The product of five more than a number and seven is thirty-five.

EXERCISE 7.6 SET B

Write each sentence as an algebraic equation. Then solve for the unknown number.

1) The sum of a number and −5 is fourteen.

2) The difference of three and a number is twelve.

3) The product of a number and −10 is one hundred.

4) The quotient of a number and twelve is two.

5) Six increased by twice a number is −18.

6) Three-fourths of a number decreased by one-fourth of the number is five.

7) The sum of a number and twenty-five is twice the number.

8) One less than two-ninths of a number is four.

9) Three times the sum of a number and three is −36.

10) One-half of a number plus ten is eleven.

11) Thirteen minus a number is three more than twice the number.

12) The product of a number and nine gives the same result as the number plus thirty-two.

13) The sum of four times a number, seven times the number, and three times the number is fifty-six.

14) The quotient of a number and six, decreased by ten, is twenty-two.

15) One-third the sum of six times a number and twelve is −20.

16) The product of three-eighths a number and four is the same as ninety less than the number.

17) Seventy-two subtracted from twice a number is equal to eighty increased by the number.

18) Three times a number added to the number is the same result as five more than the number.

19) Five times the difference of four and a number is equal to four times the number plus one.

20) The product of five less than a number and three is twenty-seven.

8.1 AN INTRODUCTION TO SETS

CHAPTER 8

Problem Place eight tennis balls in three baskets of different sizes so that there is an odd number of balls in each basket.

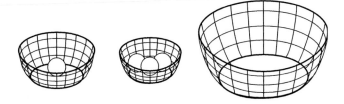

The preceding is a trick problem involving subsets. Subsets and the properties of sets are introduced in this section.

Overview For many of you, this section will be a review. However, if you have never studied sets, don't worry; all the basic definitions and concepts are presented. You will find that Venn diagrams are especially useful in work with the intersection and union of sets. Problem-solving techniques may be used to discover the number of subsets in a set with n elements.

Goals Teachers need to experience the development of mathematical language and symbolism and how these have influenced the way we communicate mathematical ideas. (*Teaching Standards*, p. 136)

Although this book gives little emphasis to the formal theory of sets, some familiarity with the notation and language of sets is useful and important. During the latter part of the 19th century, while working with mathematical entities called *infinite series*, Georg Cantor found it helpful to borrow a word from common usage to describe a mathematical idea. The word he borrowed was set.

DESCRIBING SETS

Our purposes will be served if we intuitively describe a set as a collection of objects possessing a property that enables us to determine whether a given object is in the collection. We sometimes say that a set is a *well-defined* collection—meaning that, given an object and a set, we are able to determine whether or not the object is in the set. The individual objects in a set are called **elements** of the set. They are said **to belong to** or **to be members of** or **to be in** the set. The relationship between objects in a set and the set itself is expressed in the form *is an element of* or *is a member of*.

To denote this relationship we use the following notation:

$x \in A$ means x is an **element** of set A.

$x \notin A$ means x is **not an element** of set A.

Often it is possible to specify a set by listing its members within braces. This method of describing a set is called the **tabulation** method (sometimes called the *roster method*). For example, the set of counting numbers less than 10 can be written as $\{1, 2, 3, 4, 5, 6, 7, 8, 9\}$.

A set remains the same, regardless of the order in which the elements are tabulated. Thus $\{1, 2, 3\}$ is the same set as $\{2, 1, 3\}$, $\{3, 2, 1\}$, $\{3, 1, 2\}$, $\{1, 3, 2\}$, or $\{2, 3, 1\}$. In fact, two sets are said to be **equal** if they contain exactly the same elements. $A = B$, **if and only if A and B have exactly the same elements.** If there is at least one element in either A or B that is not in the other, then $A \neq B$.

2.) Pages 215–218, 221–224, 227–230, 232–234, 238–240, 244–247, 251–261 from *Modern Mathematics,* 10th Edition by Ruric E. Wheeler and Ed R. Wheeler. © 1998 by Kendall/Hunt Publishing Company. Used with permission.

THEN AND NOW

Georg Cantor
1845-1918

An exciting feature that distinguishes mathematics from other disciplines is that mathematics deals with ideas of the infinite. Concepts of infinity plagued many mathematicians until the seminal work of Georg Cantor. We struggle with the notion of infinity because things that are infinite are beyond our realm of normal physical experience; they are therefore *abstractions* that we can deal with only in our minds. Yet Cantor showed that it makes sense to talk about the number of elements in any set, finite or infinite, and, surprisingly, showed that an *infinite* sequence of higher infinities can also be described.

Cantor was born in St. Petersburg, Russia, to Norwegian parents. The family moved to Frankfurt, Germany, when Georg was 11 years old. Resisting his father's encouragement to study engineering, Georg focused on mathematics philosophy, and physics in his studies at the Universities of Zurich, Gottingen, and Berlin. He obtained a position at the University of Halle, in Wittenberg, Germany. Although his goal was an appointment to the University of Berlin, he spent most of his life in Wittenberg.

For not attaining his goal to be appointed to the University of Berlin, Cantor blamed a Berlin mathematician named Leopold Kronecker, who did not accept any of Cantor's work on infinite sets. Kronecker openly attacked Cantor's findings, contributing to Cantor's mental collapse in 1884. This collapse was the first of many that occurred throughout the rest of his life. He died in an institution in Wittenberg in 1918, before his results were widely accepted.

In the last decade, a new chapter has been added to Cantor's work. The study of chaos theory and its geometric progeny, fractals, has renewed interest in some of Cantor's most unusual work.

Example 1

Glenda, Mark, and Martia are the only counselors in the admissions office. They constitute the set $A = \{$Glenda, Mark, Martia$\}$. Mark $\in A$; Linda $\notin A$. Can you identify other elements of set A?

Sometimes sets have so many elements that it is tedious, difficult, or even impossible to tabulate them. Sets of this nature may be indicated by a descriptive statement or a rule. The following sets are wall specified without a tabulation of members: the counting numbers less than 10, the even numbers less than 1000, the past presidents of the United States, and the football teams in Pennsylvania.

The difficulty of tabulating sets can be minimized by using **set builder notation,** which encloses within braces a letter or symbol representing an element of the set followed by a qualifying description of the element. For example, let A represent the set of counting numbers less than 10; then

$$A = \{n \mid n \text{ is a counting number less than } 10\}$$

This notation is read "the set of all elements n such that n is a counting number less than 10." Notice that the vertical line is read "such that."

Example 2

Use set-builder notation to denote the set of current United States senators.

Solution $\{x \mid x \text{ is a U.S. senator}\}$. The set is read "the set of all x such that x is a U.S. senator."

Frequently, three dots (called an **ellipsis**) are used to indicate the omission of terms. The set of even counting numbers less than 100 may be written as $\{2, 4, 6, \ldots, 98\}$. This notation saves time in tabulating elements of large sets, but it can be ambiguous unless the set has been specified completely by another description. For example, $\{2, 4, \ldots, 16\}$ could be $\{2, 4, 8, 16\}$ or it could be $\{2, 4, 6, 8, 10, 12, 14, 16\}$.

An ellipsis is also used to indicate that a sequence of elements continues indefinitely. For example, consider the set of **natural** (or **counting**) **numbers.**

$$\{1, 2, 3, 4, 5, \ldots\}$$

The set of natural numbers is an example of an **infinite set,** described informally as one that contains an unlimited number of elements. In contrast, a **finite set** contains zero elements or a number of elements that can be specified by a natural number.

> **DEFINITION: EMPTY SET**
>
> A set that contains no elements is called the *empty* or *null set* and is denoted by either $\emptyset$ or { }.

The relationship between two sets such as $A = \{1, 3, 5, 7\}$ and $B = \{1, 2, 3, 4, 5, 6, 7, 8\}$ is described by the term *subset*.

> **DEFINITION: SUBSET**
>
> Set A is said to be a *subset* of set B, denoted by $A \subseteq B$, if and only if each element of A is an element of B.

Example 3
If $A = \{x, y\}$ and $B = \{w, x, y, z\}$, then $A \subseteq B$ because each element of A is an element of B.

Example 4
If $P = \{1, 4, 7\}$ and $Q = \{4, 7, 1\}$, then $P \subseteq Q$ because each element of P is an element of Q. Moreover, $Q \subseteq P$, and $P \subseteq P$.

To show that A is not a subset of B (denoted by $A \not\subseteq B$), we must find at least one element in A that is not in B. Let's use this idea to examine whether $\emptyset$ is a subset of some set B. Using our problem-solving techniques, let's list all possibilities.

1) Either $\emptyset$ is a subset of B
2) Or $\emptyset$ is not a subset of B

In Possibility 2, if $\emptyset$ is not a subset of B, then there must be an element of $\emptyset$ not in B. But $\emptyset$ has no elements. Consequently, Possibility 2 cannot be true. The only alternative is for Possibility 1 to be true.

> **SUBSET $\emptyset$**
>
> For any set B, $\emptyset \subseteq B$.

Because all dogs are animals, the set of dogs is a subset of the set of animals. Moreover, we call the set of dogs a "proper" subset of the set of animals because some animals are not dogs.

> **DEFINITION: PROPER SUBSET**
>
> Set A is said to be a *proper subset* of set B, denoted by $A \subset B$, if and only if each element of A is an element of B and at least one element of B is not an element of A; that is, $A \subset B$ if $A \subseteq B$ and $A \neq B$.

CHAPTER 8   237

Name _____   Date _____

EXERCISE 8.1

State whether the following sets are well-defined or not well-defined.

1) {x | x is a trustworthy U.S. President}

2) {x | x is rational number between −10 and 10}

3) {x | x is a day of the week}

4) {x | x is an integer between 1 and 2}

List all the elements of each set using the tabulation method.

5) {counting numbers between 30 and 40}

6) {distinct letters in the word *happy*}

7) {x | x is a multiple of 6}

8) {odd counting numbers less than 11}

Rewrite the following sets with a verbal description using set builder notation.

9) {2, 4, 6, 8, . . ., 100}

10) {3, 6, 9, 12, 15, . . .}

11) {. . ., −3, −2, −1, 0, 1, 2, 3, . . .}

12) {January, February, . . ., December}

Classify the following sets as finite or infinite.

13) The set of counting numbers greater than one million.

14) {4, 5, 6, . . ., 14}

15) {1, 1/2, 1/4, 1/8, . . .}

16) {x | x is a fraction between 62 and 63}

17) {x | x is a biography of Abraham Lincoln in the Bloomfield College Library}

Fill in the blanks with either ∈ or ∉ so that the resulting statement is true.

18) 1950 _____ {x | x is a year in the 20th Century}

19) 5 _____ {3, 4, 5, 6}

20) −8 _____ {4, 6, 8, 10}

21) 0 _____ {−6, −3, 0, 3, 6}

22) {3} _____ {2, 3, 4, 5}

23) England _____ {x | x is a country in North America}

Fill in the blanks with either ⊆, ⊂, or ⊄ so that the resulting statement is true.

24) {Monday, Tuesday} _____ {x | x is a day of the week}

25) {Alabama, Arkansas, Alaska, Arizona} _____ {x | x is state beginning with the letter A}

26) {4, 3, 2, 1} _____ {1, 2, 3, 4} 27) {a, m, t} _____ {z, x, r, l, n, b, o, c, s}

28) ∅ _____ {−1, −2, −3, −4, −5} 29) {√5} _____ {x | x is an irrational number}

30) {10, 12, 14, 16} _____ {x | x is an even counting number}

31) $\left\{\frac{1}{2}, \frac{3}{4}, \frac{5}{6}, \frac{7}{8}, \frac{9}{10}, \frac{11}{12}\right\}$ _____ {x | x is a rational number between 0 and 1}

Let U = {2, −3, 4, −5, 6, −7, 8, −9, 10, −11, 12, −13, 14}
     A = {2, 4, 6, 8, 10, 12}   B = {−3, −5, −7, −9, −11, −13}
     C = {4, 10, −11, 12}   D = {4, 10}
     E = {6, 8, −3, −5, −7}   F = {−13, −11, −9, −7, −5}

State whether each statement is true or false.

32) A ⊂ U 33) A ⊆ B 34) D ⊆ A

35) ∅ ⊂ A 36) ∅ ⊆ ∅ 37) A ⊆ D

38) {4, 8, 10} ⊂ E 39) D = F 40) {12, −11, 4, 10} = C

41) −5 ∈ C 42) 10 ∉ B 43) x ∈ U

Choose all the sets from the right column that are subsets of the left column.

44) {s, t, u, v, w, x, y, z} (a) {3, 6}

45) {3, 6, 9, 12, 15, 18} (b) {t}

46) {x | x is a letter in the sentence *Math Is Fun*} (c) ∅

47) {..., −3, −2, −1, 0, 1, 2, 3, ...} (d) {m, a, t, s}

8.2 OPERATIONS ON SETS

If a discussion is limited to a fixed set of objects and if all elements to be discussed are contained in this set, then this overall set is called the **universal set,** or simply the **universe**. A very useful device for visualizing and discussing sets is the **Venn diagram**. Circles or other closed curves are used in Venn diagrams (named after the English logician, Robert Venn) to represent sets. The universe can be represented as the region bounded by a rectangle, and the set under consideration as the region bounded by a circle (or some other closed region) within the rectangle. In Figure 8-1, $x \in A$ means that x is some point in the circular region A. Also in Figure 8–1, set B is a proper subset of set A.

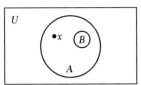

Figure 8-1

The region outside set A and inside the universe represents the complement of A, denoted by $\bar{A}$ (or sometimes by A' or $\sim A$) and read "A bar" or "bar A."

> **DEFINITION: COMPLEMENT OF A SET**
>
> The *complement* of set A, denoted by $\bar{A}$, is the set of elements in the universe that are not in set A. If A is a subset of the universe U, then $\bar{A} = \{x \mid x \in U \text{ and } x \notin A\}$.

Example 5

If $U = \{a, b, c, d\}$ and $A = \{b, c\}$, find $\bar{A}$.

Solution $\bar{A} = \{a, d\}$. Elements in U but not in A

Example 6

If the universe is all college students and if A is the set of college students who have made all A's, then all college students who have made at least one grade lower than an A is the complement of A (that is, $\bar{A}$) and is represented by the shaded region of Figure 8-2.

> **COMPLEMENT OF *A* RELATIVE TO *B***
>
> The complement of A relative to B is the set of elements in B that are not in A. This may be written as $B - A$ or in set-builder notation as
>
> $$\{x \mid x \in B \text{ and } x \notin A\}$$

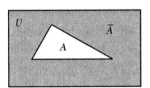

Figure 8-2

$B - A$ is sometimes read "the set difference of B and A." $B - A$, represented by the shaded region in Figure 8-3, is sometimes called a *relative complement*.

Example 7

If $A = \{x, y, z, w\}$ and $B = \{u, v, x, y\}$, then

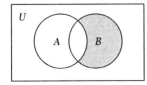

$$B - A = \{u, v\}$$
$$A - B = \{z, w\}$$
$$B - B = \emptyset$$

Figure 8-3

Special notations are used for discussing the relationships among members of two or more sets.

> **DEFINITION: INTERSECTION OF SETS**
>
> The *intersection* of any two sets A and B, denoted by $A \cap B$, is the set of all elements common to both A and B. $A \cap B = \{x \mid x \in A \text{ and } x \in B\}$.

> **DEFINITION: UNION OF SETS**
>
> The *union* of any two sets A and B, denoted by $A \cup B$, is the set of all elements in set A or in set B or in both A and B. $A \cup B = \{x \mid x \in A \text{ or } x \in B\}$.

Example 8

Let A represent a committee consisting of {Rose, Dave, Sue, John, Jack}, and let B represent a second committee consisting of {Sue, Edward, Cecil, John}. The intersection of these two sets, $A \cap B$, is {Sue, John}.

Example 9

If $A = \{1, 2, 3\}$ and $B = \{6, 7\}$, find $A \cup B$.

Solution $A \cup B = \{1, 2, 3, 6, 7\}$ Elements in either A or B

Recall that $A \cup B$ is the set of all elements that belong to A or to B or to both A and B. Any elements common to both sets are listed only once in the union. Thus, for

$$A = \{a, b, c, d, e\} \text{ and } B = \{c, d, e, f, g\} \text{ we find that}$$
$$A \cup B = \{a, b, c, d, e, f, g\}$$

Example 10

If $A = \{1, 2\}$, then $A \cup A = \{1, 2\}$.

The shaded regions in Figure 8-4 compare intersection and union under different situations for sets A and B. Notice in (a) that A and B overlap or have elements in common. In (b), A is a proper subset of B. In (c), A and B have no elements in common, so $A \cap B = \emptyset$; here A and B are said to be **disjoint**.

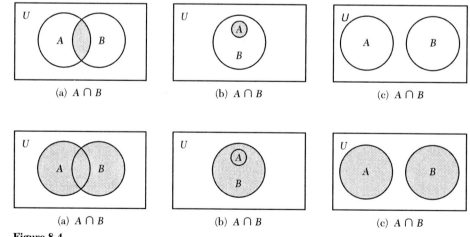

Figure 8-4

> **DEFINITION: DISJOINT SETS**
>
> Two sets A and B are said to be *disjoint* if and only if $A \cap B = \emptyset$ — that is, if the two sets have no elements in common.

Example 11

If $U = \{x \mid x \text{ is a counting number less than } 10\}$, $A = \{2, 4, 6\}$, $B = \{1, 2, 3, 4, 5\}$ and $C = \{3, 5, 7\}$, find $A \cap B, A \cap C, A \cup C,$ and $\bar{B}$.

Solution
$A \cap B = \{2, 4\}$ Elements in common
$A \cap C = \emptyset$ No elements in common
$A \cup C = \{2, 3, 4, 5, 6, 7\}$ Elements in either A or C or both
$\bar{B} = \{6, 7, 8, 9\}$ Elements in the universe, not in B

PRACTICE PROBLEM

Find the intersection and union of $A = \{1, 2, 3, \ldots, 100\}$ and $B = \{60, 61, \ldots, 1000\}$.

Answer
$A \cap B = \{60, 61, \ldots, 100\}, A \cup B = \{1, 2, 3, \ldots, 1000\}$

PROPERTIES OF SET OPERATIONS

The following properties of the intersection and union of sets can be illustrated by using Venn diagrams.

PROPERTIES OF INTERSECTION AND UNION

For all sets A, B, and C:
1) *Commutative properties*: $A \cup B = B \cup A$; $A \cap B = B \cap A$.
2) *Associative properties*: $A \cup (B \cup C) = (A \cup B) \cup C$; $A \cap (B \cap C) = (A \cap B) \cap C$
3) *Identity properties*: $A \cup \emptyset = A$; $A \cap U = A$

Notice that the *commutative* properties of intersection and union indicate that order is not important in performing these operations. The *associative* properties of intersection and union indicate that grouping is not important in performing these operations. The *identity* property of union indicates that there is a special set (the null set) with the property that its union with a set is always that same set. The universal set serves the "identity" role for intersection.

Figure 8-5(a) represents the formation of $(A \cap B) \cap C$. Similarly, Figure 8-5(b) represents the formation of $A \cap (B \cap C)$. The double-shaded region representing $(A \cap B) \cap C$ is the same as the region representing $A \cap (B \cap C)$; this illustrates the associative property, $(A \cap B) \cap C = A \cap (B \cap C)$.

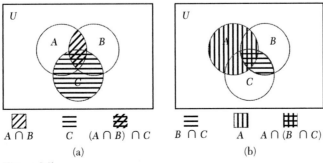

Figure 8-5

Example 12

Let $A = \{x \mid x$ is a state whose name begins with $A\}$ and $B = \{x \mid x$ is a state east of the Mississippi River$\}$. $A \cap B = \{$Alabama$\}$, and $B \cap A = \{$Alabama$\}$; therefore, $A \cap B = B \cap A$, illustrating the commutative property of intersection.

The commutative property of union is demonstrated in the next example.

Example 13

Consider $A = \{1, 3\}$ and $B = \{3, 5, 7\}$. Then $A \cup B = \{1, 3, 5, 7\}$ and $B \cup A = \{1, 3, 5, 7\}$; thus, $A \cup B = B \cup A$.

EXERCISE 8.2

The symbol (′) in the exercises refers to finding the complement of the set.

Let U = {1, 2, 3, 4, 5, 6, 7, 8, 9, 10}. Find each set.

1) Find the complement of {3, 5, 7, 9}.

2) Find the complement of {1, 10, 2, 9, 5}.

3) Find the complement of U.

4) Find the complement of ∅.

Let U = {1, 2, 3, 4, 5, 6, 7, 8, 9, 10}
    A = {2, 4, 6, 8}   B = {5, 9}   C = {2, 5, 8, 9, 10}

Find each set.

5) A ∪ B

6) A ∩ C

7) B ∩ A

8) A′

9) B′

10) C′

11) A′ ∪ B′

12) C′ ∩ B

13) (A ∩ B) ∪ C

14) A ∩ (B ∪ C)

15) A ∩ (B ∪ C)′

16) A′ ∪ (B ∩ C)

17) A − C

18) C′ ∩ ∅

19) A′ − B

Let U = {all soda pops}
    A = {all diet soda pops}   B = {all cola soda pops}
    C = {all soda pops in cans}   D = {all caffeine-free soda pops}

20) Describe, in words, the set A ∩ B.

21) Describe, in words, the set C ∪ D.

22) Describe, in words, the set A′.

23) Describe, in words, the set A − D

The following chart represents the number of athletes declaring their major.

	Baseball (B)	Soccer (S)	Tennis (T)	Total
Education (E)	24	13	19	56
Accounting (A)	30	25	18	73
Computer programming (C)	16	19	15	50
Psychology (P)	21	22	25	68
Total	91	79	77	247

Find the number of athletes in each set.

24) E ∪ P

25) C ∩ S

26) A ∪ T

27) C ∪ B

28) (B ∩ P) ∪ E

29) E′ ∩ T

Use Venn Diagrams to represent each statement.

30) A′   31) A ∪ B   32) A ∩ B

33) A − B   34) A′ ∪ B′   35) A′ ∩ B′

36) (A ∩ B) ∩ C   37) (A ∪ B) ∪ C   38) (A ∪ B) ∩ C

In a certain class, scores on a test were as follows.

Name	Score	Name	Score
Allison	45	Staci	93
Emily	89	Mindy	82
Louis	73	Harriet	79
Christopher	95	Linda	83
Roger	52	Stanley	55

Let U be the set of all students.
Let M be the set of all male students.
Let F be the set of all female students.
Let C be the set of students scoring below 80

List the elements of each set.

39) U   40) M   41) F   42) C

43) C′   44) C − M   45) F ∩ C

46) M ∪ F   47) C ∩ F′   48) (M ∩ C) ∩ F

Let U = {a, c, d, e, f, m, n, p, x, y, z}

A = {a, p, n}   B = {c, n, p, x, y, z}   C = {m, n, a, c, d, e, z}

D = {c, e, f, m, n, p, y, z}   E = {c, d, e, f, m, n, p, x, y}

Find each set.

49) A′   50) B′   51) C′   52) D′   53) E′

54) A ∩ B   55) E′ − A   56) D′ ∪ E′

57) (A − B) ∩ C′   58) (B − C′) ∩ (D ∩ E)′

59) [(D ∪ E)′ − C] ∩ (A′ ∪ B)

8.3 EQUIVALENT SETS AND CARDINAL NUMBERS

Problem A polster found that, in a group of 100 people, 40 use brand A, 25 use both brand A and brand B, and everyone uses brand A or brand B. How many people use brand B?

Overview By associating a number with the elements in a set using Venn diagrams, and utilizing problem-solving techniques, you should be able to solve the preceding problem.

The number of elements in a set provides the basis for counting. Is the number of this page a cardinal number, an ordinal number, or a nominal number? You will learn in this section the distinctions among such numbers.

Determining the number of elements in a set seems very simple. However, in probability theory (Chapter 9), we encounter some challenging problems involving the number of elements in sets. The material in this section provides the background needed to solve these problems.

Goals Construct number meanings through real-world experiences and the use of physical materials.

Develop number sense. (Standard 6, K–4)

The number concept held by mathematicians today is the result of many centuries of study. Leopold Kronecker is supposed to have said, "God created the natural numbers; all else is the work of man." His statement has been interpreted by some as meaning that the human mind is naturally endowed with the power to comprehend the concept of the counting numbers, whereas all other numbers are a result of human inventiveness.

EQUIVALENT SETS

Before discussing the concept of number, we need to understand one-to-one correspondence and equivalence.

> **DEFINITION: ONE-TO-ONE CORRESPONDENCE**
>
> A *one-to-one correspondence* between sets A and B is a pairing of the elements of A and B such that for every element a of A there corresponds exactly one element b of B, and for every element b of B there corresponds exactly one element a of A.

Example 1

Consider the sets $A = \{1, 2, 3\}$ and $B = \{\text{John, Jane, Jill}\}$. One way of placing these two sets in one-to-one correspondence is to use double-headed arrows:

$$\begin{array}{ccc} 1 & 2 & 3 \\ \updownarrow & \updownarrow & \updownarrow \\ \text{John} & \text{Jane} & \text{Jill} \end{array}$$

Of course, these two sets could be placed in one-to-one correspondence in several other ways, as well. For instance, $1 \leftrightarrow \text{Jill}$, $2 \leftrightarrow \text{John}$, and $3 \leftrightarrow \text{Jane}$.

Now try to establish a one-to-one correspondence between

$$A = \{1, 2, 3, 4\} \text{ and } B = \{\text{John, Jane, Jill}\}$$

If we let John correspond to 1, Jane to 2, and Jill to 3, no element is left to correspond to 4. Thus, a one-to-one correspondence cannot be set up between these two sets.

A number of sets can be placed in one-to-one correspondence with a given set. For example, $\{a, b, c, d\}$ can be placed in one-to-one correspondence with $\{1, 2, 3, 4\}$, and so can $\{\square, \triangle, \square, \bigcirc\}$, $\{x, y, z, w\}$, and $\{140, 180, 190, 200\}$. This characteristic leads to the following definition.

> **DEFINITION: EQUIVALENT SETS**
>
> If a one-to-one correspondence exists between two sets, the sets are said to be *equivalent* (or *matched*).

Example 2

If a room contains 70 seats, with one person sitting in each of the seats and no one standing, then the set of seats and the set of people are equivalent.

The term *equivalent* should not be confused with *equality*. If two sets are equal, each element of one set must equal a corresponding element of the other, and conversely. However, sets are equivalent if a one-to-one correspondence exists between elements. Thus, equal sets are always equivalent, but equivalent sets are not always equal.

Example 3

(a, b, c, d) and $\{b, c, a, d\}$ are equal sets; however, $\{a, b, c, d\}$ is equivalent to but not equal to $\{x, y, z, w\}$.

The relation "is equivalent to" is a relation on a collection of sets. Further, in the language of Section 2, it is reflexive (every set can be placed in one to one correspondence with itself) and symmetric (a one-to-one correspondence from A to B is also a one-to-one correspondence from B to A). With a little more thought one can see that this relation is also transitive. Thus, the relation "is equivalent to" on a collection of sets is an equivalence relation on that collection. Two sets are equivalent if they belong to the same equivalence class. This is the basis for the idea of a cardinal number discussed in the next paragraph.

CARDINAL, ORDINAL, AND NOMINAL NUMBERS

Consider four sets: $\{1, 2, 3\}$, $\{°, \#, \%\}$, $\{a, b, c\}$, and $\{\bigcirc, \triangle, \square\}$. What do these sets have in common? The answer is that they are equivalent. Intuitively, another property is probably obvious to you: they have a property of "threeness." $\{1, 2\}$, $\{a, b\}$, $\{x, y\}$, and $\{\bigcirc, \triangle\}$ are equivalent and share a property of "twoness." A property that equivalent sets have in common is called their **cardinal number**.

> **(A)**
>
> The cardinal number of set A, denoted by the symbol $n(A)$, represents the number of elements in set A. It will be read as the number of A.

This number is the cardinal number of all sets equivalent to A. Suppose that there are 30 students in a classroom. If S stands for the set of students, then $n(S) = 30$.

If $A = \{a, b, c, d\}$, than $n(A) = 4$. If $B = \{x, y, z, w\}$ then $n(B) = 4$. A is equivalent to B, and $n(A) = n(B)$. However, $A \neq B$. Just because $n(A) = n(B)$, it is not necessarily true that $A = B$. The number of a set is another set property that does not depend in any way on the kind of items represented as elements or on the order in which the elements are listed. The use of counting numbers (also called natural numbers) to tell how many objects are in a set is called *cardinal usage*. "There are 24 hours in a day" uses the counting number 24 in a cardinal sense.

In the phrase "the fifth section of the second chapter of this book" **ordinal numbers** are used. *Fifth* and *second* refer to position or order. The following example illustrates cardinal and ordinal uses of counting numbers.

Example 4

The administration building is five stories high (cardinal). The president's office is on the second floor (ordinal).

Counting numbers may also be used in a nominal sense for naming things. Social security numbers, bank account numbers, and zip codes are **nominal numbers**.

EXERCISE 8.3

State whether each of the following is a cardinal number usage or an ordinal number usage.

1) Fifth period
2) Five customers
3) Ninth inning
4) Seven stories
5) Seventh floor
6) $25

Choose all the sets from the right column that are equivalent to sets in the left column.

7) {s, t, u, v, w, x, y, z}   (a) {1, 3, 5, 7, 9, 11, 13, 15}
8) {2, 4, 6, 8, 10, 12}   (b) {t}
9) {x | x is a vowel in the word *Math*}   (c) {..., −4, −3, −2, −1, 0}
10) {0, 1, 2, 3, 4, ...}   (d) {−2, −4, −5, −7, −9, −10}
                                      (e) {Mo, Jan, Bob, Lee, Ray, Ann}
                                      (f) $\left\{\frac{1}{2}, \frac{1}{3}, \frac{1}{4}, \frac{1}{5}, \frac{1}{6}, \frac{1}{7}, \frac{1}{8}, \frac{1}{9}\right\}$

11) Let A = {1, 2, 3, 4} and B = {1, 3, 5, 7}
   a) Are sets A and B equal? Why or why not?
   b) Are sets A and B equivalent? Why or why not?

Find the cardinality of each set.

12) {2, 4, 6, 8, ..., 100}
13) {a, b, c, d, ..., x, y, z}
14) {−9, −5, 0, 2, 7, 11}
15) {Al, Bob, Carl, Dan, Evan}
16) {x | x is a letter in the word *Math*}
17) {x | x is a state of the United States}

Let U = {2, 4, 6, 8, 10, 11, 13, 15, 17, 19}
    A = {2, 6, 10, 13}   B = {11, 13, 15, 17, 19}   C = {4, 8, 10, 11, 15, 17, 19}

Find the cardinality of each set.

18) n(U)
19) n(A)
20) n(B)
21) n(C)
22) n(A′)
23) n(B′)
24) n(C′)
25) n(A ∪ B)
26) n(B ∩ C)
27) n(A − B)
28) n(A′ ∪ C)
29) n(C′ ∩ B′ ∩ A′)
30) n(A ∪ B ∪ C)

8.4 VENN DIAGRAMS AND APPLICATIONS

Example 5

A pollster found that in a group of 100 people, 40 use brand A, 25 use both brand A and brand B, and everyone uses brand A or brand B. How many people use brand B?

Can you help find the answer?

Solution To solve this problem, we use Polya's four-step method.

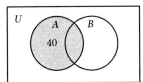

Figure 8-6

UNDERSTANDING THE PROBLEM Let's draw a Venn diagram (Figure 8-6) to visualize the given information. Now $n(A \cup B) = 100$. Why? The region bounded by the circle and designated as A represents the people who use brand A; thus, $n(A) = 40$. The region bounded by the circle and designated as B represents the people who use brand B. What does the region common to A and B represent? Is $n(A \cap B) = 25$?

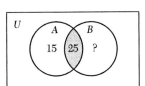

Figure 8-7

DEVISING A PLAN We can easily solve the problem by placing numbers in each closed region of Figure 3-15. Since 25 of the 40 people who use brand A also use brand B, 15 use brand A only. Now let's see if we can determine how many use brand B only.

CARRING OUT THE PLAN Because 100 people were surveyed and because everyone contacted uses either brand A or brand B, the total number in the three regions bounded by the circles of Figure 3-15 is $n(A \cup B) = 100$, or

$$15 + 25 + ? = 100$$

Thus, the number using only brand B is 60. Therefore, the total number using brand B is

$$n(B) = 25 + 60 = 85$$

LOOKING BACK In Figure 8-7 (replacing ? with 60), does the total in all regions equal 100? (Yes, $15 + 25 + 60 = 100$.) Do 40 people use brand A? (Yes, $15 + 25 = 40$.) Do 25 people use brand A and brand B? (Yes, $n(A \cap B) = 25$.)

The following discussion demonstrates the theory of the preceding example. We want to find the number of elements that are in either set A or set B, denoted by $n(A \cup B)$. The answer would not necessarily be

$$n(A) + n(B)$$

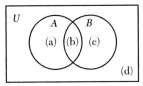

Figure 8-8

as we discover in Figure 8-8. In this figure, the number of elements in the closed regions of the Venn diagram are denoted by (a), (b), (c), and (d). Thus,

$$n(A \cup B) = (a) + (b) + (c)$$
$$n(A) = (a) + (b)$$
$$n(B) = (b) + (c)$$
$$n(A) + n(B) = (a) + (b) + (b) + (c)$$

So $n(A) + n(B)$ contains (b) more elements than $n(A \cup B)$. Because (b) $= n(A \cap B)$, we have the following theorem:

NUMBER OF ELEMENTS IN $A \cup B$

The number of elements that are in either A or B is

$$n(A \cup B) = n(A) + n(B) - n(A \cap B)$$

where $A \cap B$ consists of elements common to A and B.

Example 6

Out of 1500 freshmen, 13 students failed English, 12 students failed mathematics, and 7 students failed both English and mathematics. How many students failed English or mathematics?

Solution

$$n(E) = 13$$
$$n(M) = 12$$
$$n(E \cap M) = 7$$

To find $n(E \cup M)$,

$$n(E \cup M) = n(E) + n(M) - n(E \cap M)$$
$$= 13 + 12 - 7$$
$$= 18$$

PRACTICE PROBLEM

Forty members of Alpha Phi are enrolled in English or history. If 30 are enrolled in English and 25 are enrolled in history, how many are enrolled in both English and history?

Answer $n(E \cap H) = 15$.

NUMBER OF ELEMENTS IN $A \cup B \cup C$

For any three sets A, B, and C,

$$n(A \cup B \cup C) = n(A) + n(B) + n(C) - n(A \cap B) - n(A \cap C)$$
$$- n(B \cap C) + n(A \cap B \cap C)$$

Example 7

Out of 100 freshmen at Lance College, 60 are taking English, 50 are taking history, 30 are taking mathematics, 30 are taking both English and history, 16 are taking both English and mathematics, 10 are taking both history and mathematics, and 6 are taking all three courses. How many of the freshmen are enrolled in English, history, or mathematics?

Solution Let E represent those enrolled in English; H, those in history; and M, those in mathematics. The example states that

$$n(E) = 60 \qquad n(E \cap H) = 30$$
$$n(H) = 50 \qquad n(E \cap M) = 16$$
$$n(M) = 30 \qquad n(H \cap M) = 10$$
$$n(E \cap H \cap M) = 6$$

We are to find $n(E \cup H \cup M)$, assuming that every freshman is enrolled in one of these courses.

$$n(E \cup H \cup M) = n(E) + n(H) + n(M) - n(E \cap H) - n(E \cap M)$$
$$- n(H \cap M) + n(E \cap H \cap M)$$
$$n(E \cup H \cup M) = 60 + 50 + 30 - 30 - 16 - 10 + 6 = 90$$

JUST FOR FUN

Jennifer has 16 pieces of candy. Nine contain caramel; ten contain coconut; and six contain both caramel and coconut. Use problem-solving techniques along with sets to find out how many of the candies contain neither caramel nor coconut.

EXERCISE 8.4

Use the given information to fill in the regions of a Venn diagram with the appropriate number of elements.

1) n(U) = 50   n(A) = 38   n(B) = 32   n(A ∩ B) = 20

2) n(U) = 28   n(A) = 15   n(A ∪ B) = 18   n(A ∩ B) = 9

3) n(U) = 55   n(A′) = 27   n(B) = 23   n(A ∩ B) = 15

4) n(U) = 100   n(A) = 41   n(B) = 43   n(C) = 50
   n(A ∩ B ∩ C) = 4   n(A ∩ B) = 11   n(A ∩ C) = 16   n(B ∩ C) = 14

5) n(U) = 100   n(A) = 58   n(A ∩ B) = 36   n(A ∪ B) = 82   n(A ∩ B ∩ C) = 16
   n(A ∩ C) = 22   n(B ∩ C) = 26   n(C) = 50

6) n(U) = 150   n(A) = 64   n(A ∩ B ∩ C) = 25   n(A ∩ C) = 33   n(A ∩ B′) = 29
   n(B ∩ C) = 28   n(B ∩ C′) = 23   n(B ∪ C) = 62

7) Place the elements of the sets in the proper location on a Venn Diagram.
   U = {1, 2, 3, 4, 5, 6, 7, 8}
   A = {1, 2, 3, 4}   B = {2, 3, 6, 7}   C = {3, 4, 5, 6}

8) An entertainment magazine surveyed 185 people on whether they preferred Clint Eastwood or Spike Lee as a director.
   108 preferred Spike Lee   88 preferred Clint Eastwood   18 preferred both
   a) Draw a Venn diagram to represent the information.
   b) How many people preferred neither of the two directors?
   c) How many people preferred only Clint Eastwood as a director?
   d) How many people preferred only Spike Lee as a director?

9) A reporter for a middle school newspaper asked 105 students in the school what sport would they like to have added to their extra curricular activities—tennis or soccer.
   32 students said soccer but not tennis   42 students said tennis but not soccer
   10 students said neither sport
   a) Draw a Venn diagram to represent the information.
   b) How many students said they want both sports?
   c) How many students want only soccer or only tennis?

10) A survey of 100 randomly selected students gave the following information:
    45 students are taking math   41 students are taking English
    40 students are taking history   15 students are taking math and English
    18 students are taking math and history   17 students are taking English and history
    7 students are taking all three
    a) Draw a Venn Diagram to represent the information.
    b) How many are taking only math?
    c) How many are taking only English?
    d) How many are taking math but not history?
    e) How many are not taking any of these courses?

11) In a recent survey of 105 people, the following information was gathered:

   59 get a manicure
   35 get a facial
   19 get a manicure and a facial
   11 get a manicure, pedicure, and a facial
   51 get a pedicure
   24 get a manicure and pedicure
   13 get a pedicure and facial

   a) Draw a Venn diagram to represent the information.
   b) How many people get only a pedicure?
   c) How many people get a facial but not a manicure?
   d) How many people don't get a manicure, pedicure or facial?

12) A survey of children regarding favorite breakfast cereals produced the following results.

   44 liked Cheerios
   40 liked Rice Krispies
   16 did not like any of the three
   28 liked Rice Krispies and Cheerios
   37 liked Frosted Flakes
   18 liked all three
   23 liked Rice Krispies and Frosted Flakes
   21 liked Frosted Flakes and Cheerios

   a) Draw a Venn diagram to represent the information.
   b) How many children were surveyed?
   c) How many children liked exactly two types of cereal?
   d) How many children do not like Frosted Flakes?
   e) How many children like neither Cheerios nor Rice Krispies?

13) In the last month Lets Get Fit fitness center surveyed its members on the body part they would like to improve on. Forty members said their abdomen. Thirty-five members said their hips. Fifty members said their arms. Twelve members said their abdomen and hips. Sixteen members said their hips and arms. Twenty members said their abdomen and arms. Five members said abdomen, hips and arms. Eighteen members said none of the three.

   a) Draw a Venn diagram to represent the information.
   b) How many members took the survey?
   c) How many members wanted to improve on their abdomen only?
   d) How many members wanted to improve on only one body part?
   e) How many members wanted to work on hips or arms but not the abdomen?
   f) How many members wanted to work on hips or the abdomen?

9.1 OUTCOMES

Problem A card is dealt from a shuffled deck of 52 cards. What is the probability that the card is an ace of hearts?

Overview As this problem suggests, we shall be studying the language of uncertainty in this chapter. One significant characteristic of our increasingly complex society is that we must deal with questions for which there is no known answer but instead one or more probable (or improbable) answers. Statements like "I am 90% confident that the mean weight is between 76.2 g and 78.6 g." "This cancer treatment has a .6 chance of leading to complete remission," and "The candidate has a 20% chance of carrying the state of Ohio" are common in conversations of the day. A necessary skill for our times is the ability to measure the degree of uncertainty in an undetermined situation.

Goals Explore concepts of chance. (Standard 11, K–4)

OUTCOMES

In this section, we discuss procedures for assigning probabilities, rules that govern these probabilities, and conclusions that can legitimately be deduced from a probability once it is assigned. Because probability is a language of uncertainty, any discussion of probability presupposes a process of observation or measurement in which the outcomes are not certain. Such a process is called an **experiment**. A result of an experiment is called an **outcome**. A list of all possible outcomes in an experiment is called a **sample space**.

Example 1

Experiment: A coin is tossed.

Possible outcomes: Heads (H) or tails (T). (See Figure 9-1.)

Sample space: {H, T}.

Figure 9-1

Example 2

Experiment: A die is tossed.

Possible outcomes: The top side of the die shows 1, 2, 3, 4, 5, or 6 dots. (See Figure 9-2.)

Sample space: {1, 2, 3, 4, 5, 6}.

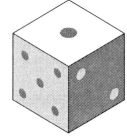

Assuming that the coin is fair, the outcomes (getting heads or getting tails) in the first experiment are said to be **equally likely,** because one outcome has the same chance of occurring as the other. Likewise, the outcomes in the second experiment (getting a 1, 2, 3, 4, 5, or 6 in the roll of a die) are equally likely.

Figure 9-2

Example 3

Suppose that we have a spinner whose needle is as likely to stop at one place as another. Suppose further that the spinner is divided into three sections labeled R, B and W. (See Figure 9-3.) Let an experiment consist of spinning the needle and observing the label of the region where it stops. A sample space for this experiment would be the list of labels:

$$\{R, W, B\}$$

Note that since the portion of the spinner labeled W is much larger than the portions of the circle labeled R or B, the outcomes of this experiment are not equally likely.

Figure 9-3

1.) Pages 009–018, 037–042, 045–047, 052–059, 062–077, 080–087, 115–127, 129–132, 141–143, 147–152, 184–187, 199–203, 206–209, from *Math for College Students: Arithmetic with Introductions to Algebra and Geometry*, 5th Edition by Ronal Straszkow. © 1997 by Kendall/Hunt Publishing Company. Used with permission.

TREE DIAGRAMS

Tree diagrams often serve to help list all possible outcomes of a multistep experiment. The procedure for drawing a tree diagram is illustrated in the following example.

Example 4

Joe must guess at both questions on a quiz. The first question is a True/False question and the second is a multiple choice question with possible answers a through d. What are the outcomes of this experiment?

Solution From a single point, draw a line to each of the possible choices on the True/False question. From each of these answers draw four lines to each possible answer on the multiple choice question. Do you see that the tree diagram in Figure 9-4 shows eight possible outcomes?

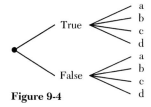

Figure 9-4

CHOOSING A SAMPLE SPACE

A first step in analyzing an experiment is the selection of a sample space. Different sample spaces can result from the same experiment, depending on how the observer chooses to record the outcomes.

Example 5

A coin is flipped twice. Find three different sample spaces.

Solution The sample spaces for this experiment can best be understood with the aid of a diagram (Figure 9-5) in which we record possible results of each flip.

Sample space A: One complete listing of the outcomes is

$$S = \{(H, H), (H, T), (T, H), (T, T)\}$$

The letter listed first in each pair indicates the result of the first flip, and the letter listed second gives the result of the second flip. Each outcome of this sample space is equally likely.

Sample space B: An alternative way to list the outcomes is to ignore the order in which the heads and tails occur and to record only how many of each appear. Thus, another possible sample space is

$$S = \{(2H), (1H \text{ and } 1T), (2T)\}$$

These outcomes are not equally likely. Because (1H and 1T) can occur in two ways (as HT and as TH), it is more likely to occur than the other two outcomes.

Sample space C: Another way to tabulate the same outcomes is to list the number of heads that occur: {0, 1, 2}. Again, the outcomes of this sample space are not equally likely. Zero heads can occur in only one way (as tails on one coin and tails on the other coin). Likewise, two heads can occur in only one way. But a result of heads on one coin and tails on the other can occur in two ways (as heads on the first coin and tails on the second, or as tails on the first flip and heads on the second).

Although the three sample spaces in the previous example are different, they share certain properties. Each set of outcomes classifies completely what can happen if the experiment is performed. Further, the members of each set of outcomes are distinct; that is, they do not overlap. In other words, each possible result of the experiment is represented by exactly one member of the set. This discussion suggests the following definition of a sample space.

> **DEFINITION: SAMPLE SPACE**
>
> A *sample space* (denoted by S) is a list of the possible outcomes of an experiment, constructed in such a way that it has the following characteristics:
> 1. The categories do not overlap.
> 2. No result is classified more than once.
> 3. The list is complete (exhausts all the possibilities).

Name _____   Date _____

EXERCISE 9.1

1) A coin is tossed in the air three times.
   a) Find the sample space of how the coin may land.
   b) How many possible outcomes are there?
   c) Are all the outcomes equally likely?
   d) In how many ways will the coin land on heads each time?

2) Mr. & Mrs. Kleese want 3 children.
   a) Find the sample space of the possible combination of genders.
   b) How many possible outcomes are there?
   c) Are all the outcomes equally likely?
   d) In how many ways can the Kleese's have all girls?

3) A six-sided green die numbered 1–6 and a six-sided red die numbered 1–6 is rolled.
   a) Find the sample space of how the dice will land. The first number is from the green die and the second number is from the red die.
   b) How many possible outcomes are there?
   c) Are all the outcomes equally likely?
   d) In how many ways can the green die and red die have a sum of 8?
   e) In how many ways is the number on the red die greater than the number on the green die?
   f) In how many ways can the green die and red die have an even sum?

4) Sean took two pairs of shoes (black and brown), two pairs of pants (black, beige), and six shirts (blue, black, green, beige, white, and grey) on a trip. Assume all items are compatible.
   a) List all the possible outcomes of wearing an outfit.
   b) How many possible outcomes are there?

5) The Salad-To-Go Deli offers a variety of salads. There are 3 choices for lettuce: iceberg, romaine, or Boston. There are 4 choices for toppings: tomatoes, cucumbers, bacon bits, or croutons. There are 2 choices for dressing: Italian or Blue cheese. *(you can only make a salad with one choice from each category. i.e. you cannot make a salad with romaine, tomatoes, croutons and Italian dressing.)*
   a) List all the possible outcomes of making a salad.
   b) How many different types of salads are offered?
   c) How many types of salads could you make if you want to use iceberg lettuce?
   d) How many types of salads could you make if you want cucumbers and Italian dressing?
   e) How many types of salads could you make if the Deli had no more bacon bits or romaine lettuce?

6) On the lunch menu The Casa Mia Restaurant offers appetizers (bruschetta or fried calamari), soups (minestrone), salads (house or Caesar), entrées (lasagna or eggplant parmagiana) and desserts (tiramisu or cannoli). The lunch special is an appetizer, a soup, a salad, an entreè, and a dessert for $13.99.
   a) List all the possible outcomes if a person orders one item from each category.
   b) How many possible outcomes are there for lunch?
   c) How many possible outcomes are there if you know you want lasagna as your entrée as part of the lunch special?
   d) How many possible outcomes for the lunch special are there if the restaurant ran out of fried calamari for the appetizer and tiramisu for the dessert?

7) Roger is making his class schedule for next semester, which must include one class from each of the four categories; an English class {Medieval Literature or Modern Poetry}; a Mathematics class {College Algebra or Math for Liberal Arts}; a Computer Science class {Intro to Spreadsheets or Web Page Design} and a Sociology class {Aging in America, Minorities in America, or Women in American Culture}.

   a) List all the possible class schedules Roger can have.

   b) How many different schedules can Roger create if all classes shown are available?

   c) How many different schedules can Roger create if he is not eligible for College Algebra and Intro to Spreadsheets?

   d) How many different schedules can Roger create if all sections of Minorities in America, Modern Poetry and Aging in America are closed?

   e) How many different schedules can Roger create if he needs to take College Algebra and Medevil Literature to fulfill his degree requirements?

8) The following members of the Lets Save the Earth club who are eligible to be elected for an office are Arnold, Bob, Cindy, Don, and Ellen. *(No person can hold two offices.)*

   a) List all the possible outcomes if a president and vice-president need to be elected.

   b) List all the possible outcomes if Don is president, and a vice-president and secretary need to be elected.

   c) List all the possible outcomes if the president elected is a woman, the vice-president elected is a man, and the treasurer is a woman.

9) At the Economist Club annual benefit dinner, the elected officers and honored guests are seated at a straight head table. The elected officers are Mr. Town (president), Mrs. Rocca (vice-president) and Mr. Chang (treasurer). The honored guests are Mr. Diaz, Mrs. Abdu, and Mrs. McCann.

   a) List all the possible seating arrangements of the head table if the president sits in the first end seat, the treasurer sits in the third seat, and the vice-president sits at the other end seat.

   b) List all the possible seating arrangements of the head table if Mr. Chang, Mr. Diaz, and Mrs. Abdu sit next to each other with Mrs. Abdu sitting between Mr. Chang (on her left) and Mr. Diaz (on her right).

   c) List all the possible seating arrangements of the head table if the men and women sit in alternate seats and Mrs. McCann must sit in the last end seat.

10) At a recent university retirement luncheon, the two honored professors came with their spouses. The president of the university sat in the middle seat at the straight head table.

   a) List all the possible seating arrangements of the head table if men are to be seated on one side of the president and women on the other side of president.

   b) List all the possible seating arrangements of the head table if men and women are to be seated alternately.

   c) List all the possible seating arrangements of the head table if husband and wife are to be seated alternately.

   d) List all the possible seating arrangements if the husband must sit to the left of his wife.

11) Andy, Betty, Clyde, Dawn, Evan and Felicia have reserved six seats in a row at the theatre starting at an aisle seat and ending at the wall.

   a) List all the possible seating arrangements if Clyde must be seated on the aisle seat and Dawn must sit next to him.

   b) List all the possible seating arrangements if Betty must sit on the aisle seat with Andy sitting next to her and Felicia must sit in the seat next to the wall.

   c) List all the possible seating arrangements if the men all sit together (Evan is sitting in the aisle seat) and all the women sit together.

12) A two-digit number will be written using the digits 0, 1, 2, 3, and 4.
    a) List all the possible two-digit numbers that can be written using the given digits
    b) List all the possible two-digit numbers if the given digits cannot be repeated.
    c) List all the possible two-digit numbers that are an even number.
    d) List all the possible two-digit numbers that are a multiple of five.
13) A three-digit number will be written using the digits 0, 2, 4, 7, and 9.
    a) List all the possible three-digit numbers that can be written using the given digits.
    b) List all the possible three-digit numbers if the given digits cannot be repeated.
    c) List all the possible three-digit numbers that are a multiple of four.
    d) List all the possible three-digit number that are an odd number and no digit is repeated.
14) Employees will need a four-letter code to enter the laboratory. The code will be created using the letters of a, m, e, and x.
    a) List all the possible four-letter codes that can be created if a letter cannot be repeated.
    b) List all the possible four-letter codes that can be created if the first letter and last letter must be vowels and letter may be repeated.
    c) List all the possible four-letter codes that can be created if the consonants must be next to each other and no letter may be repeated.
15) Each cash-box lock has three-letter codes using the letters p, o, and t.
    a) List all the possible three-letter codes that can be created.
    b) List all the possible three-letter codes that can be created if a letter is not repeated.
    c) List all the possible three-letter codes that can be created if the last letter must be consonant.
    d) List all the possible three-letter codes that can be created if the middle letter must be a vowel and no letter may be repeated.
16) A computer password requires a code with four digits and/or letters. The password will be created using e, m, 3, and 8.
    a) List all the possible codes that can be created with a digit or letter and no repetitions.
    b) List all the possible codes in which the first two are letters and the last two are digits and no letter or digit is repeated.
    c) List all the possible codes in which a letter and digit alternate starting with a letter and no digit or letter is repeated.
    d) List all the possible codes in which the even digit is first and the vowel is last and no digit or letter is repeated.

9.2 THE FUNDAMENTAL PRINCIPLE OF COUNTING

Problem The Hardy College Bulldogs are purchasing uniforms. Members can purchase red or white shorts. They can choose red, white, or striped shirts. How many possible choices are there for the uniforms?

Overview In a uniform sample space the probability of an event is computed by counting the outcomes in the event and the outcomes in the sample space and then dividing. In some cases it is easy to list all the outcomes and then count them. In other cases, it is difficult to list all the outcomes. Hence, we need to learn some procedures that will allow us to count sets without listing all of their elements.

Goals Other topics that should be introduced include ... elementary counting techniques, ...

As we have seen earlier, tree diagrams are useful in listing all the possible outcomes of many experiments. Observe carefully the tree diagrams in the next two examples, and you may discover a hint about a powerful counting technique.

Example 1

The college chorale is planning a concert tour with performances in Dallas, St. Louis, and New Orleans. In how many ways can its itinerary be arranged?

Solution If there is no restriction on the order of the performances, any one of the three cities can be chosen as the first stop. After the first city is selected, either of the other cities can be second, and the remaining city is then the last stop. (See Figure 9-6.)

Did you notice that there were three choices for the first stop, two choices for the second stop, and one choice for the third stop for a total of $3 \cdot 2 \cdot 1 = 6$ possible itineraries?

Example 2

The members of the chorale in Example 1 decided to sing first in New Orleans, next in Dallas, and finally in St. Louis. Now, they must decide on their modes of transportation. They can travel from the campus to New Orleans by bus or plane; from New Orleans to Dallas by bus, plane, or train; and from Dallas to St. Louis by bus or train. The tree diagram in Figure 9-7 indicates the chorale's options. The first part of the trip can be made in 2 ways, the second part in 3 ways, and the last part in 2 ways. Notice that the total number of ways the transportation can be chosen is $2 \cdot 3 \cdot 2 = 12$ ways.

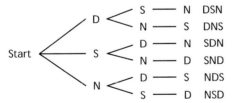

Figure 9-6

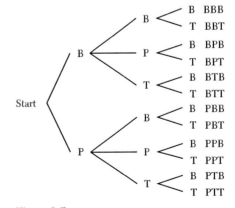

Figure 9-7

These examples suggest the following principle, called the *Fundamental Principle of Counting*.

> **FUNDAMENTAL PRINCIPLE OF COUNTING**
>
> 1. If an experiment consists of two steps, performed in order, with n_1 possible outcomes of the first step and n_2 possible outcomes of the second step, then there are
>
> $$n_1 n_2$$
>
> possible outcomes of the experiment.
>
> 2. In general, if k steps are performed in order, with possible number of outcomes $n_1, n_2, n_3, \ldots n$ respectively, then there are
>
> $$n_1, n_2, n_3, \ldots n_4$$
>
> possible outcomes of the experiment.

Example 3

A coin is tossed 5 times. If we classify each outcome as either a head or a tail, how many outcomes are in the sample space?

Solution Because there are 5 steps, each with 2 possible outcomes, there are $2 \cdot 2 \cdot 2 \cdot 2 \cdot 2 = 32$ different outcomes.

PRACTICE PROBLEM

A die is rolled 3 times. How many different outcomes are there?

Answer 216

The Fundamental Principle of Counting is helpful in solving problems such as the following.

Example 4

In the state of Georgia, automobile license plates contain an arrangement of 3 letters followed by 3 digits. If all letters and digits may be used repeatedly, how many different arrangements are available?

Solution There are 26 letters to choose from for each of the 3 letter places, and there are 10 digits to choose from for the digit places. By the Fundamental Principle of Counting, the number of arrangements is

$$26 \cdot 26 \cdot 26 \cdot 10 \cdot 10 \cdot 10 = 17{,}576{,}000$$

Example 5

If the letters and numbers on a license plate in Georgia (see Example 4) are assigned randomly, what is the probability that you will receive a license plate on which the letters read *DOG*?

Solution In Example 4 we learned that the sample space contains 17,576,000 outcomes, all of which are equally likely. By the counting principle there are

$$1 \cdot 1 \cdot 1 \cdot 10 \cdot 10 \cdot 10 = 1000$$

possible license plates that begin with *DOG* (because we have only 1 choice for each of the letters). Thus the probability of a license plate with the word *DOG* is

$$\frac{1000}{17{,}576{,}000} = \frac{1}{17{,}576}$$

Example 6

An urn contains 5 red balls and 7 white balls. A ball is drawn, its color is noted, but the ball is not replaced. A second ball is drawn. What is the probability of drawing a red ball followed by a white ball?

Solution 1 By the counting principle, there are 12 ways of drawing the first ball and 11 ways of drawing the second. Therefore, there are

$$12 \cdot 11 = 132$$

ways of drawing the 2 balls. All of these ways are equally likely. At the same time, there are 5 ways of drawing the red ball on the first draw and 7 ways of drawing the white ball on the second draw. Therefore, the number of ways of drawing a red ball and then a white ball is

$$5 \cdot 7 = 35$$

Thus,
$$P(R \text{ followed by } W) = \frac{35}{132}$$

Solution 2 We, of course, recognize that this problem could have been solved with a tree diagram as discussed in the previous section. Consider the tree diagram in Figure 9-8 and then observe that

$$P(R \text{ followed by } W) = \frac{5}{12} \cdot \frac{7}{11} = \frac{35}{132}$$

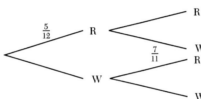

Figure 9-8

Example 7

In how many ways can 6 students line up outside Wheeler's office to complain about their grades?

Solution As a first step, we can choose a person to be first in line; there are 6 ways to do this. Then we can choose a second person; only 5 persons are available after the first person is chosen. Similarly, the third place must be filled by one of 4 persons and so on.

First	Second	Third	Fourth	Fifth	Sixth
6	5	4	3	2	1

By the Fundamental Principle of Counting, there are $6 \cdot 5 \cdot 4 \cdot 3 \cdot 2 \cdot 1$ or 720 ways to accomplish this task.

Name _____   Date _____

EXERCISE 9.2

1) Mr. & Mrs. Van Saunn plan to have 3 children.
   a) List all the possible outcomes of the combination of genders.
   b) How many possible outcomes are there to have 3 children?
2) How many possible outcomes are there for the coin to land if it is tossed six times?
3) In how many ways can a sandwich be made if there are 4 choices of deli meat, 3 choices for cheese, 2 choices for toppings, and 4 choices for a spread *(one item per category)*?
4) Sam doesn't know any of the answers for a twelve-question true or false test, so he must guess on all the questions. How many different ways can he mark his answer sheet?
5) The letters of the word "PUNCH" are rearranged to create a 5-letter code.
   a) How many codes can be created if each letter can be repeated?
   b) How many codes can be created if a letter cannot be repeated?
   c) How many codes can be created if the first letter must be a vowel and repetition of letters is allowed?
   d) How many codes can be created if the last letter must be a consonant and no repetition of letters is allowed?
6) A two-digit number can be formed using the digits of 0, 1, 2, 3, 4, 5, 6, 7, 8, and 9.
   a) How many two-digit numbers can be formed if repetition of digits is allowed?
   b) How many two-digit numbers can be formed if no repetition of digits is allowed?
   c) How many two-digit numbers can be formed if the first number must be a four and repetition of digits is not allowed?
   d) How many two-digit numbers can be formed if the number is an odd number and repetition of digits is allowed?
   e) How many two-digit numbers can be formed if the last digit is a five or a two and repetition of digits is not allowed?
7) A three-digit number can be formed using the digits of 0, 1, 2, 3, 4, 5, 6, 7, 8, and 9.
   a) How many three-digit numbers can be formed if repetition of digits is allowed?
   b) How many three-digit numbers can be formed if no repetition of digits is allowed?
   c) How many three-digit numbers can be formed if the first number must be a nine and repetition of digits is allowed?
   d) How many three-digit numbers can be formed if the number is an even number and repetition of digits is allowed?
   e) How many three-digit numbers can be formed if the first digit is a five and the last digit is a two and repetition of digits is not allowed?
8) A four-digit number can be formed using the digits of 0, 1, 2, 3, 4, 5, 6, 7, 8, and 9.
   a) How many four-digit numbers can be formed if repetition of digits is allowed?
   b) How many four-digit numbers can be formed if no repetition of digits is allowed?
   c) How many four-digit numbers can be formed if the last number must be a multiple of three and repetition of digits is not allowed?
   d) How many four-digit numbers can be formed if the number is an odd number and repetition of digits is not allowed?
   e) How many four-digit numbers can be formed if the first two digits must be a six or a seven and repetition of digits is not allowed?

9) A 4-digit number greater than 3000 can be formed using the digits 3, 4, 5, 6, 7, 8, and 9.
   a) How many 4-digit numbers can be created if repetition of numbers is allowed?
   b) How many 4-digit numbers can be created if repetition of numbers is not allowed?
   c) How many 4-digit numbers can be created if the first number must be even and repetition of numbers is not allowed?
   d) How many 4-digit numbers can be created if the last number must be a multiple of 4 and repetition of numbers is allowed?

10) Roger, Isabella, Herbert, and Amanda are eligible to run for office in the student government. *(No person can hold two offices.)*
    a) In how many ways can a president and a vice-president be elected?
    b) In how many ways can a president, vice-president and treasurer be elected?
    c) In how many ways can a president and vice-president be elected if the president must be a woman?
    d) In how many ways can a president, vice-president, and treasurer be elected if the treasurer is Roger?

11) In a certain state all domestic animals must have a coded identification tag. The id tags are of the following type: the first place must be an 8, the second and third place must be letters and the fourth and fifth place can be any number. If repetition of letters and numbers is allowed, how many different identification tags can be made?

12) There are 7 members on the Board of Directors of the Ross Chemical Corp. as follows: Alice Johnson, Steve Bachtel, George Klagman, Jeremy Billingsley, Shawn Washington, Wallace McKenzie and Colleen Brooks. At any stockholder meeting these members are seated at a straight head table.
    a) In how many different ways can the Board members be seated?
    b) Jeremy and Wallace are bitter enemies and cannot be seated together. Therefore, they are seated in the last seat on each end. How many different seating arrangements are possible?
    c) How many number of ways can the Board members be seated if the two women must be seated next to each other?

13) At a recent corporation retirement luncheon, five employees came with their spouses and are seated at the head straight table.
    a) How many seating arrangements are possible if there are no restrictions?
    b) How many seating arrangements are possible if men and women are to be seated alternately?
    c) How many seating arrangements are possible if husband and wife are to be seated alternately?
    d) How many seating arrangements are possible if a husband sits to the left of the wife?
    e) If the CEO of the corporation sits in the middle seat, how many seating arrangements are possible if men are to be seated on one side of the CEO and women on the other side of the CEO?

14) In some foreign countries telephone numbers consist of five digits.
    a) How many telephone numbers can be created if there are no restrictions?
    b) How many telephone numbers can be created if no digit can be repeated?
    c) How many telephone numbers can be created if the first number must be a 3 and repetitions of numbers are allowed?

15) Tadishi's ZIP code is 95841. How many different ZIP codes can be formed using all of those same digits?

16) How many different results are possible if five coins are tossed?

17) How many different license "numbers" can a state issue if each includes two letters followed by four digits and the first and last digits cannot be zeros (repetition of letters and digits is allowed)?

18) George keeps four textbooks and three novels on his desk between bookends.
    a) In how many ways can he arrange the books if he does not care how they are arranged?
    b) In how many ways can he arrange the books if he wants the textbooks grouped together?
    c) In how many ways can he arrange the books if he wants all the novels to be to the left of the textbooks?

19) Andy, Betty, Clyde, Edward, Felicia, and Gloria have reserved six seats in a row at the theater, starting at an aisle seat and ending at the wall.
   a) In how many ways can they arrange themselves if there are no restrictions?
   b) In how many ways can they arrange themselves if Clyde must sit in the aisle seat?
   c) In how many ways can they arrange themselves if Andy and Betty must sit next to each other?
   d) In how many ways can they arrange themselves if the men will sit together and the women sit together?

20) In "Super Lotto" you select six distinct numbers from the counting numbers 1 through 51, hoping your selection will be the winner.
   a) How many different sets of six numbers can you select?
   b) Diane always includes her age and her husband's age as two of her selections. How many different sets of six numbers can she select?
   c) Greg always chooses odd numbers as his selections. How many different sets of six numbers can he select?

21) Subject identification numbers in a certain scientific research project contains three letters followed by three digits and then three more letters. Assume repetitions are not allowed within any of the three groups, but letters in the first group of three may occur also in the last group of three. How many distinct identification numbers are possible?

22) Carole has eight errands to run today, five of them pleasant, but the other three unpleasant.
   a) How many ways can she plan her day if she decides to put off the unpleasant errands for another day?
   b) How many ways can she plan her day if she is determined to complete all eight errands today in no particular order?
   c) How many ways can she plan her day if she will complete all her unpleasant errands first?
   d) How many ways can she plan her day if she will complete all her pleasant errands together?

23) Use the set {0, 1, 2, 3, 4, 5, 6, 7, 8, 9, a, b, c, d, e, f, u, v, w, x} to answer the following questions.
   a) How many four-digit numbers are possible if the first number must be a six?
   b) How many license plates can be created if three letters come first followed by four digits with no repetition of letters and digits?
   c) How many different ways are there to create a secret code if all the elements of the set have to be used and a letter and a digit alternate starting with a letter?

24) How many different seven-digit phone numbers are there if the first two digits cannot be ones or zeros and repetition of numbers is allowed?

25) How many three-digit numbers can be written using the digits 0, 1, 2, 3, 4, 5 if repetition of digits are allowed?

26) There are five boys and six girls at a birthday party.
   a) Find the number of ways of arranging them in a line for ice cream.
   b) Find the number of ways of arranging them in two separate lines, girls in one line, boys in another.
   c) Find the number of ways of arranging them in one line if the girls must be together and the boys must be together.

9.3 PROBABILITY OF EVENTS

To each outcome of a sample space we can assign a number, called its **probability,** that measures how likely that outcome is to occur. The probability of an outcome is a number between 0 and 1 inclusive. If the probability of an outcome is near 1, that outcome will likely occur when the experiment is performed. If the probability of an outcome is near 0, it is unlikely that that outcome will occur when the experiment is performed. A probability of zero indicates there is no outcome, and a probability of 1 indicates the outcome is the entire sample space.

PROPERTIES OF PROBABILITY

A **probability assignment on a sample space** must satisfy two properties:

1. If A is a possible outcome, then its probability $P(A)$ is between 0 and 1 inclusive.
2. The sum of the probabilities of all outcomes in the sample space equals 1.

For some sample spaces we are able to assign probabilities to outcomes based on our understanding of the properties of the sample space. Such assignments are called **theoretical probabilities.**

Example 6

In the experiment of tossing a fair coin the outcomes of a head and a tail are equally likely to occur. Hence we will assign probabilities $P(H) = \frac{1}{2}$ and $P(T) = \frac{1}{2}$. Note that these assignments satisfy the properties of probability:

(a) $0 \leq P(T) \leq 1$ and $0 \leq P(H) \leq 1$

(b) $P(S) = P(H) + P(T) = \frac{1}{2} + \frac{1}{2} = 1$

Example 7

Consider the experiment of spinning the spinner in Figure 9-3. A sample space for this experiment would be $\{R, B, W\}$. If we believe that the spinner is twice as likely to stop on region W as on regions R or B, then we would assign a probability of $\frac{1}{2}$ to outcome W and probabilities of $\frac{1}{4}$ to outcomes R and B. Again,

(a) $0 \leq P(R) \leq 1, 0 \leq P(W) \leq 1, 0 \leq P(B) \leq 1$

(b) $P(R) + P(W) + P(B) = 1$

UNIFORM SAMPLE SPACE

There is one whole class of sample spaces, called *uniform sample spaces,* whose probability assignments are particularly easy to determine.

DEFINITION: UNIFORM SAMPLE SPACE

If each possible outcome of the sample space is equally likely to occur, the sample space is called a *uniform sample space.*

Suppose that a uniform sample space consists of m possible outcomes $\{A_1, A_2, \ldots, A_m\}$. Because each of the outcomes is equally likely, it seems reasonable to assign to each outcome A_i the same probability, denoted by P. Because the sum of the probabilities of the m individual outcomes must be 1, it follows that

$$P(A_1) + P(A_2) + P(A_3) + \ldots + P(A_m) = 1$$
$$\underbrace{P + P + P + \ldots + P}_{m \text{ times}} = 1$$
$$mP = 1 \quad \text{or} \quad P = \frac{1}{m}$$

Thus, each of the m outcomes has probability $1/m$.

> **EQUAL PROBABILITIES**
>
> In a uniform sample space with m possible outcomes, each outcome has probability $1/m$. This is sometimes written as
>
> $$P(A) = \frac{1}{n(S)} = \frac{1}{m}$$
>
> where A represents one outcome and $n(S)$ represents the number of possible outcomes in the sample space.

Example 8

Eight identical balls numbered 1 to 8 are placed in a box. (See Figure 9-9.) Find a sample space and probability assignments describing the experiment of randomly drawing one of them from the box.

Solution A suitable sample space is $\{1, 2, 3, 4, 5, 6, 7, 8\}$, each number representing one of the 8 balls. Because each ball is equally likely to be drawn, we assign a probability of $\frac{1}{8}$ to each outcome:

$$P(1) = \frac{1}{8}, \quad P(2) = \frac{1}{8}, \quad \ldots, P(8) = \frac{1}{8}$$

Example 9

A card is drawn from a shuffled deck (52 cards with four suits: clubs, diamonds, hearts, and spades; each suit with 13 cards: 2–10, jack, queen, king, and ace). What is the probability of drawing the ace of hearts? (See Figure 9-10.)

Figure 9-9

Figure 9-10

Solution A uniform sample space for this experiment would consist of a listing of all 52 cards. Hence each outcome would be assigned probability 1/52. In particular, the probability of drawing the ace of hearts is 1/52.

EMPIRICAL PROBABILITY

An alternative way to assign probabilities to outcomes in an experiment involves performing the experiment many times and looking at the empirical data. Consider the following example.

Example 10

A fair die is rolled 10,000 times. Table 9-1 itemizes the number of times a 1 has occurred at various stages of the process.

Table 9-1

Number of Rolls (N)	Number of 1's Occuring (m)	Relative Frequency (m/N)
10	4	.4
100	20	.2
1000	175	.175
3000	499	.166333
5000	840	.168
7000	1150	.164285714
10,000	1657	.1657

Notice that, as N (number of rolls) becomes larger, the relative frequency (m/N) stabilizes in the neighborhood of 0.166 ≈ 1/6. Thus, we are willing to assign the probability

$$P(1) = \frac{1}{6}$$

Compare this thinking with our previous method of assigning probabilities. Assume that the die is constructed so that the outcomes are equally likely. Because the sample space $S = \{1, 2, 3, 4, 5, 6\}$ is a uniform sample space, the probability of obtaining a 1 in a sample space of 6 outcomes is 1/6, the same answer we obtained by looking at the empirical data.

In the previous example, we assigned probability to an outcome by assigning the fraction of times that the outcome occurred when the experiment was performed a large number of times. Similarly, suppose that a thumbtack lands with its point up 1000 times out of 10,000 trials. The relative frequency is $\frac{1000}{10,000} = \frac{1}{10}$. If we repeat the experiment 10,000 more times and find that the ratio is still approximately $\frac{1000}{10,000} = \frac{1}{10}$, we will agree to assign this number as a measure of our degree of belief that it will land point up on the next toss. These examples suggest the following definition.

> **DEFINITION: EMPIRICAL PROBABILITY**
>
> If an experiment is performed N times, where N is a very large number, the probability of an outcome A should be approximately equal to the following ratio:
>
> $$P(A) = \frac{\text{number of times } A \text{ occurs}}{N}$$

Example 11

A loaded die (one for which the outcomes are not equally likely) is thrown a number of times with results as shown in Table 9-2.

(a) How many possible outcomes are there?

(b) What are the possible outcomes?

(c) Using the frequency table (Table 9-2), assign a probability to each of the outcomes.

Table 9-2

Outcome	1	2	3	4	5	6
Frequency	967	843	931	1504	1576	1179

> **THEN AND NOW**
>
> **BLAISE PASCAL**
> **1623–1662**
>
> Blaise Pascal was a frail youth whose mother died when he was 3 and whose upbringing fell to his father, Etienne, a French lawyer and mathematician. Blaise and his two sisters were educated at home by their father, and Blaise never attended a school or university.
>
> His attraction to mathematics was evident by the time he was 12. At age 16, he had already published an original treatise on conic sections, a profound 1-page geometric result known today as Pascal's Theorem. Such a promising first paper showed indication of a brilliant career in geometry, but Pascal quickly turned to other mathematical interests. At age 18, he embarked on the unusual enterprise of designing, building, and selling mechanical calculating machines. Pascal's friend, the Chevalier de Mèrè, posed a gambling problem, whereupon Pascal—with Pierre de Fermat—originated the modern mathematical theory of probability. In this work, Pascal made use of the array that follows. Though this array was known by JiǎXiàn in China in the 11th century, it is known as Pascal's Triangle because it was Pascal who developed its properties and showed its connections with other facets of mathematics.
>
> Pascal's name first appears these days in our mathematics curriculum by way of Pascal's Triangle, usually in grades 7 and 8. Students are asked to write down the numbers in the next row by looking for the pattern in the first few rows. Then they are asked to find other patterns within Pascal's Triangle, or to find the sum of each row, and so on. Then, throughout the first 2 or 3 courses of algebra, Pascal's work reappears in various contexts.
>
> ```
>               1
>             1   1
>           1   2   1
>         1   3   3   1
>       1   4   6   4   1
>     1   5  10  10   5   1
>   1   6  15  20  15   6   1
> ```
>
> We owe to Pascal appreciation for his innovation in the science of computation (IBM has one of his mechanical calculators in its museum). A popular and widely used computer language is named PASCAL. A great deal of attention is now being paid to the subject of artificial intelligence, and Pascal pondered the question of whether machines could think three centuries ago.

Solution There are 6 possible outcomes to this experiment: The die may show a 1, 2, 3, 4, 5, or 6. By adding the frequencies of the 6 outcomes, we see that the experiment was performed 7000 times. Thus the relative frequency of the outcome of 1 is 967/7000 or approximately 0.14. Similarly, the relative frequency of a 2 is 843/7000 or approximately 0.12. Continuing in this way, we make the following probability assignments:

$P(1) = .14 \qquad P(2) = .12 \qquad P(3) = .13$

$P(4) = .21 \qquad P(5) = .23 \qquad P(6) = .17$

PROBABILITY OF EVENTS AND PROPERTIES OF PROBABILITY

Problem (a) Two coins are tossed, What is the probability of getting at least one head?

(b) Of the freshmen who entered Loren College last year, 12% made A's in freshman English 8% made A's in history, and 4% made A's in both English and history. An admissions counselor would like to know what percent made A's in English or history.

Overview In the previous section we computed probabilities of a single outcome from an experiment. In this section we will examine the problem of computing the probability of an event where an event is a subset of outcomes from the sample space. Because an event is a subset of the sample space, we will find that many of the concepts and much of the language of Chapter 8 on sets will be useful. In particular, we will renew our acquaintance with the words end and or and the notion of complement of a subset.

Goals Make predictions based on experimental or theoretical probabilities (Standard 11, 5-8).

We now extend our definition of probability to events.

> **DEFINITION: EVENT**
>
> An *event* is a subset of a sample space.

Example 1

In the previous section we found a uniform sample space for the experiment of tossing a pair of coins. Tabulate the event "at least one coin shows a head."

Solution

Sample Space	Event
(HH, HT, TH, TT)	(HH, HT, TH)

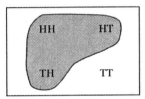

Figure 9-11

The event consists of the possible outcomes in the sample space that include at least one occurrence of heads, circled in Figure 9-11.

In the case of an event from a uniform sample space, the probability of an event is particularly easy to compute.

> **DEFINITION: PROBABILITY OF AN EVENT**
>
> Let S represent a uniform sample space with $n(S)$ equally likely outcomes, and let A be an event in S. If A is an event consisting of $n(A)$ outcomes, then $P(A)$ is given by
>
> $$P(A) = \frac{n(A)}{n(S)}$$

This rule is the classical definition of probability. Suppose that there are N equally likely possible outcomes of an experiment. If r of these outcomes have a particular characteristic so that they can be classified as a success, then the probability of a success is defined to be r/N.

Example 2

Suppose that a card is drawn from a set of 6 (numbered 1 through 6). There are 6 equally likely possible outcomes of the experiment. Let 2 of these, a 3 and a 6, represent a success E. Then

$$P(E) = \frac{n(E)}{n(S)} = \frac{2}{6} = \frac{1}{3}$$

PRACTICE PROBLEM

What is the probability of drawing an ace from a standard deck?

Answer   4/52 or 1/13

Example 3

Return to the experiment of tossing a pair of coins. What is the probability of tossing

(a) at least one head?

(b) exactly one head?

Solution   Using the uniform sample space
$$S = \{HH, HT, TH, TT\}$$
we can tabulate the events in question.

(a) $A = \{HH, HT, TH\}$ (b) $B = \{HT, TH\}$

Thus,

(a) $P(A) = \dfrac{n(A)}{n(S)} = \dfrac{3}{4}$ (b) $P(B) = \dfrac{n(B)}{n(S)} = \dfrac{1}{2}$

Example 4

A poll is taken of 500 workers to determine whether they want to go on strike. Table 9-3 indicates the results of this poll.

Table 9-3

In Favor of a Strike	Against a Strike	No Opinion
280	200	20

(a) What is the probability that a worker selected at random is in favor of a strike?

(b) What is the probability that such a worker has no opinion?

Solution   One sample space for this experiment would consist of a list of the 500 workers. To choose a worker *at random* means that each employee has the same chance of being selected; hence the sample space is uniform.

(a) Let E be the event "worker is in favor of the strike." Then
$$P(E) = \frac{n(E)}{n(S)} = \frac{280}{500} = \frac{14}{25}$$

(b) Similarly, $P(\text{no opinion}) = \dfrac{20}{500} = \dfrac{1}{25}$

Because events are subsets, we can, of course, form unions, intersections, and complements of events. The properties that govern unions, intersections, and complements of events are very important.

DEFINITION: AND, OR, AND COMPLEMENT

1. The event $A \cup B$ (A or B) is the collection of all outcomes that are in A or in B or in both A and B.
2. The event $A \cap B$ (A and B) is the collection of all outcomes that are in both A and B.
3. The **complement** of an event A, denoted by $\bar{A}$, is the collection of all outcomes that are in the sample space and are not in A.

Example 5

In the rolling of a fair die, what is the probability that the result will be an odd number or a 4?

Solution We let O represent an odd number and F represent a 4, and then we seek $P(O \cup F)$. In Figure 9-12(b), using Venn diagrams, we see that

$$P(O \cup F) = \frac{n(O \cup F)}{n(S)} = \frac{4}{6} = \frac{2}{3}$$

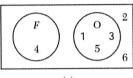

Notice in Figure 9-12(a) that

$$P(O) = \frac{3}{6} \quad \text{and} \quad P(F) = \frac{1}{6}$$

Furthermore, $P(O \cup F) = P(O) + P(F)$

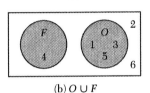

because $\frac{4}{6} = \frac{3}{6} + \frac{1}{6}$

(b) $O \cup F$

Figure 9-12

Example 6

In the rolling of the same fair die, what is the probability of getting either an even number or a multiple of 3?

Solution Let E represent an even number, and let M represent a multiple of 3. We seek $P(E \cup M)$. In Figure 9-13(b) we see that

$$P(E \cup M) = \frac{n(E \cup M)}{n(S)} = \frac{4}{6} = \frac{2}{3}$$

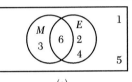

Notice in Figure 9-13(a) that

$$P(E) = \frac{3}{6} \quad \text{and} \quad P(M) = \frac{2}{6}$$

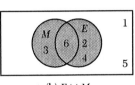

Here, however,

$$P(E \cup M) \neq P(E) + P(M)$$

because $\frac{4}{6} \neq \frac{3}{6} + \frac{2}{6}$

(b) $E \cup M$

Figure 9-13

What is the difference between the problems in the two previous examples? For $P(O \cup F)$, F and O had no points in common. For $P(E \cup M)$, E and M overlapped. This discussion suggests the following definition and property of probability.

MUTUALLY EXCLUSIVE EVENTS

1. Events A and B are *mutually exclusive* if they have no outcomes in common.
2. If events A and B are mutually exclusive,

$$P(A \cup B) = P(A) + P(B)$$

Example 7

From a standard deck of cards, you draw one card. What is the probability of your getting a spade or a red card?

Solution Verify that $P(S) = \frac{13}{52}$ and $P(R) = \frac{26}{52}$

Because the outcome of getting a spade and the outcome of getting a red card are mutually exclusive,

$$P(S \cup R) = P(S) + P(R)$$
$$= \frac{13}{52} + \frac{26}{52} = \frac{39}{52} = \frac{3}{4}$$

Now let's return to Example 6, where we noted that $P(E \cup M) = \frac{4}{6}$, $P(E) = \frac{3}{6}$, and $P(M) = \frac{2}{6}$. The reason $P(E \cup M) \neq P(E) + P(M)$ is that the outcome 6 is in both E and M and thus is counted twice in $P(E) + P(M)$. The probability of getting an outcome that is in both E and M is

$$P(E \cap M)$$

Because $E \cap M$ is included twice in $P(E) + P(M)$, we subtract one of these and note that

$$P(E \cup M) = P(E) + P(M) - P(E \cap M)$$

or

$$\frac{4}{6} = \frac{3}{6} + \frac{2}{6} - \frac{1}{6}$$

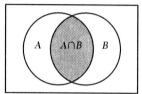

We can generalize this concept by realizing that in set theory the number of outcomes in event A or in event B is the number in A plus the number in B less the number in $A \cap B$, which has been counted in both A and B. (See Figure 9-14.) Thus,

$$n(A \cup B) = n(A) + n(B) - n(A \cap B)$$

Figure 9-14

In a uniform sample space we can divide both sides of the equation by N, the number of elements in a sample space, to obtain probabilities

$$\frac{n(A \cup B)}{N} = \frac{n(A)}{N} + \frac{n(B)}{N} - \frac{n(A \cap B)}{N}$$

Thus,
$$P(A \cup B) = P(A) + P(B) - P(A \cap B)$$

PROBABILITY OF *A* OR *B*

For any two events A and B, the *probability of A or B* is given by

$$P(A \cup B) = P(A) + P(B) - P(A \cap B)$$

We return now to the second of our introductory problems.

Example 8

Of the freshmen who entered Loren College last year, 12% made A's in English, 8% made A's in history, and 4% made A's in both English and history. What percent made A's in English or history?

Solution
$$P(E) = .12$$
$$P(H) = .08$$
$$P(E \cap H) = .04$$
$$P(E \cup H) = P(E) + P(H) - P(E \cap H)$$
$$= .12 + .08 - .04$$
$$= .16$$

16% made A's in English or history.

Example 9

In drawing a card from 8 cards numbered 1 through 8, what is the probability of getting an even number or a number less than 5?

Solution   Let A represent the event of getting a number less than 5, and let B represent the event of getting an even number. Then

$$P(A) = \frac{4}{8} = \frac{1}{2}$$
$$P(B) = \frac{4}{8} = \frac{1}{2}$$

But 2 and 4 are both even and less than 5; consequently,

$$P(A \cap B) = \frac{2}{8} = \frac{1}{4}$$

Therefore,

$$P(A \cup B) = P(A) + P(B) - P(A \cap B)$$
$$= \frac{1}{2} + \frac{1}{2} - \frac{1}{4} = \frac{3}{4}$$

PRACTICE PROBLEM

A card is drawn from a standard deck of cards. What is the probability that it is either an ace or a spade?

Answer $\frac{4}{13}$

The complement of event A is everything in the sample space except A denoted by $\bar{A}$. Because $A \cup \bar{A} = S$, and $A \cap \bar{A} = \emptyset$, $P(A \cup \bar{A}) = 1$ and $P(A \cap \bar{A}) = 0$, then $P(A \cup \bar{A}) = P(A) + P(\bar{A}) = 1$. Therefore, $P(\bar{A}) = 1 - P(A)$.

PROBABILITY OF THE COMPLEMENT OF A

If $\bar{A}$ is the complement of A, then

$$P(\bar{A}A) = 1 - P(A).$$

Example 10

What is the probability of not getting an ace in the drawing of a card from a standard deck of cards?

Solution $P(\text{no ace}) = 1 - P(\text{ace}) = 1 - \frac{4}{52} = \frac{12}{13}$

THINGS TO KNOW ABOUT A "STANDARD" DECK OF CARDS

- A standard deck of cards contains 52 cards.
- The deck consists of into two colors: 26 red cards (diamonds & hearts)
                                           26 black cards (spades & clubs)

- There are 4 suits:

	hearts (♥)	diamonds (♦)	clubs (♣)	spades (♠)
Each suit has 13 cards:	2	2	2	2
	3	3	3	3
	4	4	4	4
	5	5	5	5
	6	6	6	6
	7	7	7	7
	8	8	8	8
	9	9	9	9
	10	10	10	10
	J	J	J	J
	Q	Q	Q	Q
	K	K	K	K
	A	A	A	A

- ACES can have a value of 1 (low) or 11 (high). You will have to be told if the ace will be low or high.
- Face cards are the Jacks, Queens and Kings.
- Face cards do **not** have a number value. For example, when you are asked for cards greater than 8, your answer will only be 9 and 10. A jack is not greater than a 10.

THINGS TO KNOW ABOUT ROLLING TWO DICE

- There are a total of 36 ways you can roll two dice.
- A double is rolling the same number, e.g. (2,2) is a double.

The following table will break down how many ways you can roll the dice to the *sums*.

SUM OF:	2	3	4	5	6	7	8	9	10	11	12
	(1,1)	(1,2)	(1,3)	(1,4)	(1,5)	(1,6)	(2,6)	(3,6)	(4,6)	(5,6)	(6,6)
		(2,1)	(2,2)	(2,3)	(2,4)	(2,5)	(3,5)	(4,5)	(5,5)	(6,5)	
			(3,1)	(3,2)	(3,3)	(3,4)	(4,4)	(5,4)	(6,4)		
				(4,1)	(4,2)	(4,3)	(5,3)	(6,3)			
					(5,1)	(5,2)	(6,2)				
						(6,1)					

EXERCISE 9.3

1) There is a 65% probability that it will rain tomorrow. What is the probability that it will not rain tomorrow?

2) An entertainment newsletter stated there was an 88% probability that Star Wars: Episode III would be the top grossing movie of the year. What is the probability that another movie will be the top grossing movie of the year?

3) The probability that a high school senior smokes is $\frac{3}{5}$. What is the probability that a high school senior does not smoke?

4) The probability of Richard bowling two perfect games in one year is 0.02. What is the probability that Richard will not bowl two perfect games in one year?

5) There are eight balls in a box numbered one through eight. Is the event of choosing a "seven" ball or a "five" ball from the box mutually exclusive?

6) Greg, Marcia, Peter, Jan, Bobby, Cindy, DJ, Stephanie, Michelle, Willis, Arnold, and Kimberly are all in the toddler's group at a daycare. Is the event of Stephanie or Bobby being chosen to lead in the recital of the days of the week mutually exclusive?

7) There are twelve balls in a box. Six balls are red and numbered one through six and the other six balls are green and numbered one through six. Is the event of choosing a red ball or a ball numbered three mutually exclusive?

8) In a standard 52-card deck, is the event of drawing a five or a six mutually exclusive?

9) In a standard 52-card deck, is the event of drawing a heart or a ten mutually exclusive?

10) In a standard deck of cards, is the event of drawing a face card or a card greater than seven mutually exclusive?

11) In a true/false question, what is the probability that you will answer the question correctly?

12) In a multiple choice question with 5 choices, what is the probability that you will answer the question incorrectly?

13) A company surveyed its employees on whether they wanted Fridays off in June, July, and August or the week between Christmas and New Years off. Ninety-five employees wanted the Fridays off and one hundred fifteen wanted the week off. If an employee is selected at random, what is the probability that the employee wanted Fridays off?

14) John has sixteen pairs of socks in his drawer — 4 gray pairs, 4 black pairs, 4 blue pairs, and 4 white pairs.
    a) If a pair of socks is pulled from the drawer, will each pair have an equally likely chance of being pulled?
    b) What is the probability that John will choose a black pair of socks?
    c) Is the event of choosing a blue pair or a gray pair mutually exclusive?
    d) What is the probability that John will choose a blue pair of socks or a gray pair of socks?

15) The Youth Social Club of a local town will hold elections for the office of president. Sabrina, Joan, Debbie and Anna each have an equally likely chance of winning. What is the probability that Anna will be elected president?

16) Mr. & Mrs. Teller plan to have four children.
    a) List all the possible outcomes of the genders of the four children.
    b) How many possible outcomes are there for the Tellers to have all boys?
    c) What is the probability that the Tellers will have all boys?
    d) How many possible outcomes are there for the Tellers to have one girl?
    e) What is the probability that the Tellers will have one girl?
    f) What is the probability that the Tellers will have at least one boy?

17) Vibhuti tossed a quarter, a dime, a nickel, and a penny, respectively, in the air.
    a) List all the possible outcomes of how the coins land.
    b) How many possible outcomes are there of the quarter and dime landing on heads?
    c) What is the probability of the quarter and dime landing on heads?
    d) How many possible outcomes are there of all the coins landing on tails?
    e) What is the probability of all the coins landing on tails?
    f) What is the probability of at least two coins landing on heads?
    g) What is the probability that none of the coins will land on heads?

18) A green and red six-sided die, both numbered one through six, are rolled.
    a) List all the possible outcomes of how the dice land.
    b) What is the probability of rolling the sum of eight?
    c) What is the probability of rolling the sum of one?
    d) What is the probability of rolling the sum of four?
    e) What is the probability of the green die being greater than the red die?
    f) What is the probability of rolling a double?
    g) What is the probability of rolling the sum of an even number?
    h) What is the probability of not rolling the sum of ten?

19) A card is drawn from a standard 52-card deck.
    a) What is the probability of drawing a king?
    b) What is the probability of drawing a four?
    c) What is the probability of drawing a black card?
    d) What is the probability of drawing a spades?
    e) What is the probability of drawing an ace of hearts?
    f) What is the probability of drawing a black three?
    g) What is the probability of drawing a face card?
    h) What is the probability of drawing a card greater than seven (aces are high)?
    i) What is the probability of drawing an odd numbered card?
    j) What is the probability of not drawing a ten?

20) The letters of the sentence, *Math is Fun*, are each written on an index card and placed into a bag. A card is chosen from the bag.
    a) What is the probability of choosing a card with the letter f?
    b) What is the probability of choosing a card with a vowel on it?
    c) What is the probability of not choosing a card with the letter m?
    d) What is the probability of choosing a card with the letter t or the letter n?
    e) What is the probability of choosing a card with the letter h or a vowel?
    f) What is the probability of choosing card with the letter a or choosing a card with a consonant?

21) Refer to the previous problem of rolling the green and red die.
    a) What is the probability of rolling a sum of five or a sum of nine?
    b) What is the probability of rolling a sum of seven or a double?
    c) What is the probability of rolling a sum of six or the green die showing a four?
    d) What is the probability of rolling a sum of four or a double?
    e) What is the probability of rolling the sum of an odd number or the sum of one?
    f) What is the probability of rolling the sum of the multiple of four or the green die being a larger number than the red die?

g) What is the probability of not rolling the sum of five or rolling the sum of twelve?
h) What is the probability of not rolling the sum of six or rolling a double?

22) A card is drawn from a standard 52-card deck.
   a) What is the probability of drawing a two or a ten?
   b) What is the probability of drawing a face card or a five?
   c) What is the probability of drawing a diamond or a club?
   d) What is the probability of drawing a four of clubs or a three of diamonds?
   e) What is the probability of drawing a card less than four (aces high) or a queen?
   f) What is the probability of drawing a red card or a king?
   g) What is the probability of drawing a spade or an ace?
   h) What is the probability of drawing an even numbered card or a black two?
   i) What is the probability of drawing a card greater than eight (aces low) or a jack of spades?
   j) What is the probability of drawing a heart or a three?
   k) What is the probability of drawing a black card or a six?
   l) What is the probability of not drawing a nine or not drawing a three?
   m) What is the probability of not drawing a diamond or drawing a red queen?

The following chart represents the distribution of readers and the books they read.

	Mystery	History	Romance	Total
Women (under 30 y/o)	45	11	63	119
Women (between 30 and 55)	34	25	75	134
Women (over 55)	56	15	74	145
Total	135	51	212	398

y/o refers to "years old"

23) Based on the above data, find the probability of each of the following events.
   a) the reader was under 30 y/o.
   b) the reader read history books.
   c) the reader was between 30 and 55 y/o or read mystery books.
   d) the reader was a 50 y/o male who read romance books.
   e) the reader was a female over 55 y/o who read history books.
   f) the reader did not read a mystery book.
   g) the reader read history books or romance books.
   h) the reader read mystery books or was under 30 years old.

SOLUTIONS TO EXERCISES

Exercise 1.1 Set A
1) false
2) false
3) false
4) false
5) true
6) true
7) true
8) false
9) true
10) true
11) {−5, 8, 0, 55}
12) {8, 0, 5}
13) {−5, 8, 2.7, $-\frac{4}{5}$, $-\frac{2}{5}$, 0, 0.25, 55, 2.58}
14) { $\sqrt{7}$, $-\sqrt{3}$, π, 0.2835843...}
15) entire set of numbers

Exercise 1.1 Set B
1) true
2) true
3) false
4) false
5) true
6) false
7) true
8) true
9) false
10) false
11) {0, $\sqrt{81}$}
12) {−78, 0, $\sqrt{81}$}
13) {−5.3434..., −78, $\frac{1}{3}$, 6.50, 0, −0.11, $\sqrt{81}$, $3\frac{4}{5}$}
14) {$\sqrt{8}$, π, $\frac{\sqrt{6}}{5}$}
15) entire set of numbers

Exercise 1.2
1) eight hundred fifty-two
2) four thousand, two hundred fifty-six
3) seventeen thousand, one hundred nine
4) three million, fifty-seven thousand, ten
5) fourteen million, one hundred thousand, seven hundred
6) nine hundred forty-six thousand, three
7) one billion, three hundred fifty-seven million, nine hundred twenty-six thousand, one hundred eighty-three
8) 745
9) 50,068
10) 105,006
11) 40,000,036
12) 5,007,238
13) 12,015,000,000
14) 89,000,000,089
15) 313,710,000
16) 712,000,422
17) 5 millions
18) 9 tens
19) 8 hundred thousands
20) 6 thousands
21) 7 ten thousands

Exercise 1.3
1) 50
2) 80
3) 30
4) 510
5) 1470
6) 5900
7) 7000
8) 500
9) 900
10) 700
11) 14,700
12) 27,900
13) 180,000
14) 1,538,300
15) 215,749,540
16) 215,749,500
17) 215,750,000
18) 215,750,000
19) 215,700,000
20) 216,000,000
21) 220,000,000
22) 200,000,000

Exercise 1.4
1) prime
2) prime
3) composite
4) composite
5) composite
6) composite
7) 2×3^2
8) $2^3 \times 3$
9) 2×23
10) 2×17
11) $2 \times 3 \times 7$
12) $2 \times 3 \times 11$
13) $2^2 \times 5^3 \times 11$
14) $2^2 \times 5^3 \times 7$
15) 3×5^2
16) 5×7^2
17) $5^3 \times 11$
18) $5^3 \times 13$
19) $2^2 \times 3 \times 5 \times 11$
20) $2 \times 5^2 \times 13$
21) $2 \times 3 \times 37$
22) $3^2 \times 43$
23) 7×19
24) 11×13

Exercise 1.5

1) 6896
2) 25708
3) 34497
4) 46976
5) 2555
6) 6744
7) 1446
8) 6946
9) 288
10) 13860
11) 31240
12) 30603
13) 560000
14) 12000
15) 325
16) 19290
17) 180
18) 651
19) 9659
20) 456
21) 785
22) 359875
23) 15
24) 63350
25) 22529
26) 4356
27) 935
28) 1157
29) 27105
30) 69136
31) 434
32) 894783

Exercise 1.6 Set A

1) 21
2) 8
3) 94
4) 39
5) 35
6) 40
7) 48
8) 115
9) 10
10) 45
11) 15
12) 96
13) 93
14) 0
15) 18
16) 20
17) 241
18) 6
19) 30
20) 117
21) 17
22) 3

Exercise 1.6 Set B

1) 50
2) 64
3) 49
4) 61
5) 63
6) 115
7) 144
8) 176
9) 100
10) 63
11) 15
12) 144
13) 86
14) 17
15) 36
16) 213
17) 100
18) 14
19) 75
20) 155
21) 0
22) 2

Exercise 1.7 Set A

1a) Obstetricians-Gynecologists, Cardiologists, Neurologist, Plastic Surgeon
1b) 104
1c) 71
1d) $123,530
1e) $23,608
2) 17
3) $1095
4) $95
5a) $458
5b) $6844
5c) yes
6a) $2845
6b) $1863
6c) yes
7a) $8933
7b) $4448
7c) yes
8) $360
9) $225
10a) $95
10b) $204
10c) yes
11) $1114
12a) $2354
12b) $2888
12c) $517
12d) $611
13) 3150
14) 912,500
15) 2290
16) yes
17) 24
18) $585
19) $240
20a) $486
20b) 3
20c) $531
21) group rate
22) $12.25
23) $4.00
24) $0.75
25a) $0.17
25b) $0.16
25c) store #2
26) $18
27) $866
28) $3600
29) $305
30) $1715
31a) $9900
31b) $2475
32a) $5000
32b) $15000
33a) $100
33b) $1800
34) $375
35a) 11
35b) $88
36a) 1225
36b) 49
37) $725,000
38) 2190
39) 72
40a) 12
40b) 40
40c) $512
41) 195
42) 51
43) $691,130
44) $2350
45) $289
46) 353
47) $3000
48) $6
49) no

Exercise 1.7 Set B

1a) 60
1b) 6
1c) 2
1d) 4
2) $300
3a) 104
3b) 61
3c) 13
4a) $5745
4b) $2652
5a) $28750
5b) $19716
5c) $9034
6a) $1230
6b) $1099
6c) no
7a) $1195
7b) $620
8) $240
9) $205
10) $3756
11) $103
12) $845
13) yes
14) 128
15) 104
16) less
17) 610
18a) 1750
18b) 2400
18c) yes
19) 96
20) $4040
21) $84,845
22) no
23) $10
24) $410
25) $0.30
26) $40
27a) $0.60
27b) $0.55
27c) $82.50
28) $295
29) $35
30) $116
31) $2975
32a) $26,250
32b) $6562.50
33a) $4200
33b) $3600
33c) $50
34a) $330
34b) $27.50
34c) $6.35
34d) $0.91
35) 14,400
36) $102
37) $733
38) 3041
39) 8
40) $4676
41) 560
42) 625
43) 3116
44) $1671
45) 130
46) $3162
47) $188
48) 6759
49) 39
50) $7.5 billion or $7,500,000,000
51a) $2500
51b) $870

Exercise 2.1 Set A

1) 40
2) 10
3) −40
4) −10
5) −10
6) 10
7) −8
8) −35
9) −14
10) −77
11) −30
12) −16
13) −74
14) −182
15) −13
16) 1
17) −6
18) −1
19) −32
20) 2
21) 0
22) −104
23) −32
24) 0
25) 28
26) 28
27) −4
28) 4
29) 78
30) −24
31) 0
32) 23
33) 50
34) −70
35) −5
36) 3

Exercise 2.1 Set B

1) 14
2) −8
3) −28
4) −1
5) 7
6) 8
7) −4
8) −42
9) −18
10) −88
11) −99
12) −16
13) −106
14) −166
15) −39
16) −9
17) −26
18) −3
19) −31
20) −14
21) 27
22) −89
23) −128
24) 31
25) 42
26) 33
27) −6
28) −22
29) 18
30) −32
31) 40
32) 34
33) 10
34) −32
35) 5
36) −1

Exercise 2.2

1) 15
2) 24
3) −21
4) −18
5) −32
6) −30
7) 48
8) 35
9) 72
10) 63
11) 4
12) 9
13) −8
14) −6
15) −7
16) −7
17) 8
18) 4
19) 5
20) 6
21) −13
22) −12

Exercise 2.3 Set A

1) −109
2) 8
3) −8
4) −19
5) −35
6) 40
7) 48
8) −185
9) 12
10) −58
11) −7
12) 24
13) 115
14) −12
15) −42
16) −280
17) −61
18) −10
19) 78
20) 109
21) 37
22) −1

Exercise 2.3 Set B

1) −160
2) 32
3) −11
4) −49
5) 63
6) 16
7) −576
8) 112
9) −1
10) −61
11) 35
12) −324
13) −46
14) −21
15) −12
16) 178
17) 24
18) 58
19) 66
20) 135
21) −37
22) −2

Exercise 2.4 Set A

1) $x + 6$
2) $3 + x$
3) $2x - 5$
4) $2x - 6$
5) $6x$
6) $9x$
7) $3x + 4$
8) $5x + 2$
9) $\dfrac{x}{7}$
10) $\dfrac{x}{4}$
11) $2x + 6 = 24$
12) $x - 2 = 16$
13) $2(x + 6) = 24$
14) $4x + 7 = -13$
15) $-7x - 5 = 86$
16) $3x - 10 = x + 6$
17) $2x + 3x + 4x = 90$
18) $3(x - 5) = -25 + 4$
19) $4(x + 3) = 48$
20) $2(x + 3) + x = x + 18$

Exercise 2.4 Set B

1) $x + 2$
2) $8 + x$
3) $2x - 7$
4) $2x - 4$
5) $5x$
6) $3x$
7) $4x + 6$
8) $3x + 1$
9) $\dfrac{x}{9}$
10) $\dfrac{x}{2}$
11) $2x + 4 = 18$
12) $x - 5 = 12$
13) $2(x + 3) = 32$
14) $5x + 9 = 39$
15) $-6x - 5 = 67$
16) $3x - 4 = x + 12$
17) $2x + 3x + 5x = 90$
18) $4(x - 6) = -21 + 5$
19) $3(x + 4) = 18$
20) $2(x + 5) + x = x + 16$

Exercise 3.1 Set A

1) true
2) true
3) true
4) true
5) true
6) true
7) true
8) false
9) true
10) true
11) $\dfrac{1}{2}, \dfrac{5}{8}$
12) $\dfrac{8}{5}, \dfrac{8}{8}, \dfrac{7}{3}, \dfrac{8}{1}$
13) $4\dfrac{2}{3}, 2\dfrac{1}{2}$
14) $\dfrac{77}{100}$
15) $\dfrac{31}{55}$
16) $\dfrac{174}{511}$
17) $\dfrac{697}{2000}$
18a) $\dfrac{111}{726}$
18b) $\dfrac{564}{726}$
19a) 610
19b) $\dfrac{151}{610}$
19c) $\dfrac{217}{610}$
19d) $\dfrac{393}{610}$
20) $\dfrac{971}{1000}$

Exercise 3.1 Set B
1) 8
2) 7
3) 5
4) 4
5) $5\frac{2}{3}$
6) $4\frac{1}{6}$
7) $1\frac{3}{7}$
8) $1\frac{3}{8}$
9) $7\frac{2}{5}$
10) $9\frac{2}{5}$
11) $7\frac{13}{16}$
12) $8\frac{4}{17}$
13) $65\frac{2}{7}$
14) $56\frac{4}{7}$
15) $\frac{9}{1}$
16) $\frac{8}{1}$
17) $\frac{3}{2}$
18) $\frac{8}{5}$
19) $\frac{21}{8}$
20) $\frac{20}{7}$
21) $\frac{37}{4}$
22) $\frac{25}{3}$
23) $\frac{75}{2}$
24) $\frac{313}{6}$
25) $\frac{65}{6}$
26) $\frac{188}{9}$
27) $\frac{871}{7}$
28) $\frac{629}{3}$
29) $\frac{780}{17}$
30) $\frac{627}{32}$

Exercise 3.2
1) 3
2) 4
3) 6
4) 9
5) 28
6) 32
7) 60
8) 36
9) 30
10) 12
11) $\frac{1}{5}$
12) $\frac{1}{3}$
13) $\frac{5}{9}$
14) $\frac{2}{3}$
15) $\frac{7}{11}$
16) $\frac{9}{13}$
17) $\frac{2}{3}$
18) $\frac{3}{8}$
19) $\frac{5}{12}$
20) $\frac{11}{15}$
21) $\frac{26}{147}$
22) $\frac{28}{117}$
23) $\frac{3}{4}$
24) $\frac{13}{17}$
25) $\frac{2}{5}$
26) $\frac{3}{4}$
27) $\frac{4}{7}$
28) $\frac{6}{7}$

Exercise 3.3
1) $\frac{10}{27}$
2) $\frac{27}{40}$
3) $\frac{2}{9}$
4) $\frac{2}{7}$
5) $\frac{8}{15}$
6) $\frac{1}{12}$
7) 12
8) $13\frac{1}{2}$
9) 6
10) 0
11) $\frac{1}{2}$
12) $1\frac{3}{4}$
13) $1\frac{4}{5}$
14) $1\frac{3}{5}$
15) $\frac{9}{55}$
16) 35
17) 0
18) undefined
19) $\frac{1}{5}$
20) $\frac{3}{5}$
21) $3\frac{3}{14}$
22) 5
23) $19\frac{1}{4}$
24) 117

25) $1\frac{1}{15}$
26) $21\frac{1}{3}$
27) $3\frac{15}{28}$
28) $1\frac{36}{49}$
29) $4\frac{3}{4}$
30) 18
31) 4
32) $\frac{1}{3}$
33) $10\frac{13}{48}$
34) $14\frac{1}{16}$
35) $19\frac{17}{21}$
36) undefined

Exercise 3.4

1) $\frac{11}{12}$
2) $1\frac{7}{20}$
3) $\frac{6}{25}$
4) $\frac{3}{11}$
5) $1\frac{1}{8}$
6) $1\frac{17}{30}$
7) $1\frac{1}{6}$
8) $2\frac{5}{21}$
9) $\frac{2}{21}$
10) $\frac{11}{30}$
11) $\frac{2}{15}$
12) $1\frac{11}{12}$
13) $6\frac{6}{7}$
14) $7\frac{7}{8}$
15) $8\frac{2}{5}$
16) $4\frac{3}{5}$
17) $13\frac{9}{10}$
18) $13\frac{1}{60}$
19) $30\frac{23}{30}$
20) $35\frac{9}{20}$
21) $3\frac{7}{8}$
22) $4\frac{11}{60}$
23) $\frac{6}{7}$
24) 9
25) $19\frac{1}{9}$
26) $14\frac{6}{11}$
27) $2\frac{5}{6}$
28) $2\frac{45}{56}$
29) $6\frac{3}{4}$
30) $13\frac{3}{16}$
31) $3\frac{8}{9}$
32) $26\frac{2}{3}$
33) $1\frac{1}{15}$
34) 144
35) $4\frac{2}{3}$
36) $1\frac{3}{10}$

Exercise 3.5 Set A

1) $\frac{25}{36}$
2) $1\frac{1}{4}$
3) $\frac{1}{72}$
4) $\frac{1}{9}$
5) 1
6) $14\frac{1}{36}$
7) $72\frac{1}{2}$
8) $5\frac{2}{21}$
9) 2
10) $\frac{1}{88}$
11) $12\frac{1}{6}$
12) $2\frac{1}{2}$

Exercise 3.5 Set B

1) $\frac{9}{64}$
2) $1\frac{1}{6}$
3) $\frac{1}{800}$
4) $\frac{16}{81}$
5) 3
6) $21\frac{7}{8}$
7) 83
8) $7\frac{11}{45}$
9) $1\frac{5}{6}$
10) $\frac{17}{33}$
11) $13\frac{1}{6}$
12) $2\frac{5}{12}$

Exercise 3.6 Set A

1) $\dfrac{90}{140} = \dfrac{9}{14}$

2) $\dfrac{200}{350} = \dfrac{4}{7}$

3) $\dfrac{10}{35} = \dfrac{2}{7}$

4a) $\dfrac{300}{1200} = \dfrac{1}{4}$

4b) $\dfrac{75}{400} = \dfrac{3}{16}$

4c) $1860

5) $\dfrac{4}{10}$

6a) A: $3\dfrac{14}{21}$ B: $3\dfrac{18}{21}$

6b) Machine A

7a) Bag #1

7b) Bag #2 and Bag #3

8) $10\dfrac{1}{2}$

9) no

10a) $36\dfrac{11}{12}$

10b) $48\dfrac{1}{4}$

10c) no

11) $9\dfrac{1}{20}$

12a) 11

12b) $55\dfrac{1}{2}$

13) $7\dfrac{1}{2}$

14) $7\dfrac{19}{20}$

15a) $11\dfrac{1}{6}$

15b) 11

15c) $14\dfrac{23}{30}$

16a) $2\dfrac{1}{8}$

16b) $9\dfrac{3}{8}$

17) $13\dfrac{1}{4}$

18) 51

19) $11

20) $20\dfrac{5}{6}$

21) 3

22a) 1830

22b) $24\dfrac{2}{5}$

22c) $30

23) $41\dfrac{1}{2}$

24) 40

25) 195

26) 12

27) $16\dfrac{2}{3}$

28) $5668

29) 1011

30) 576

31) 1024

32) $4\dfrac{13}{18}$

33) $348

34) $32\dfrac{6}{7}$

35a) $46.96

35b) $140.88

36a) 890

36b) $84.55

37) $33\dfrac{3}{4}$

38) 99

39a) $146\dfrac{2}{3}$

39b) $47.67

40) $95\dfrac{7}{10}$

41a) $95\dfrac{15}{32}$

41b) $34.37

42a) 9

42b) $113.25

43a) $293.50

43b) $126.50

44) $1395

45) $\dfrac{1}{2}$

46a) $25\dfrac{1}{5}$

46b) $1638

47) 16

48) $2\dfrac{1}{2}$

49) 14

Exercise 3.6 Set B

1) $\dfrac{30}{48} = \dfrac{5}{8}$

2) $\dfrac{11}{40} = \dfrac{55}{200}$

3a) $\dfrac{25}{129} = \dfrac{75}{387}$

3b) $\dfrac{12}{87}$

3c) $54,856

4) $\dfrac{12}{16}$

5) $3\dfrac{70}{100}$

6a) Oak: $\dfrac{48}{20}$ Pine: $\dfrac{55}{20}$

6b) Oak

7a) #1: $\dfrac{129}{4}$ #2: $\dfrac{259}{8}$

7b) Bottle #2

7c) Bottle #1 only

8) 14

9) yes

10a) $7\dfrac{17}{20}$

10b) $7\dfrac{7}{8}$

10c) yes

11a) $21\dfrac{1}{2}$

11b) $3\dfrac{5}{9}$

12) yes

13a) $8\dfrac{11}{12}$

13b) $12\dfrac{7}{60}$

13c) $15\dfrac{11}{30}$

14a) $1\dfrac{1}{4}$

14b) $\dfrac{1}{2}$

14c) $2050

15a) $5\dfrac{9}{10}$

15b) $2\dfrac{7}{30}$

16) 822

17) $9\dfrac{3}{5}$

18) 7

19) $\dfrac{23}{40}$

20) $64\dfrac{1}{2}$

21) $173.25

22) $56\dfrac{1}{4}$

23) 9

24) 75

25) $414

26) $2559.38

27) $26\dfrac{2}{3}$

28) $5\dfrac{1}{2}$

29) 204

30) 9

31) $1152

32) $123

33a) 1500

33b) $108

34) 66

35a) $74\dfrac{4}{5}$

35b) $19.26

36a) 232

36b) $154\dfrac{2}{3}$

36c) $77\dfrac{1}{3}$

37) $31\dfrac{1}{6}$

38a) $66\dfrac{1}{4}$

38b) $35.35

39) $3.63

40a) $13\dfrac{3}{4}$

40b) $110

41a) $262

41b) $78

42) 12,000

43) 384

44) $2817.25

45) $26\dfrac{1}{4}$

46) $2\dfrac{5}{6}$

47) 1066

48) $240.63

49) 624

50a) $3\dfrac{3}{4}$

50b) $26.25

51) 25

Exercise 3.7 Set A

1) $-\dfrac{125}{216}$
2) $-1\dfrac{1}{4}$
3) $-\dfrac{1}{72}$
4) $-\dfrac{1}{9}$
5) -1
6) $11\dfrac{11}{16}$
7) $27\dfrac{1}{2}$
8) $3\dfrac{17}{35}$
9) 1
10) $-\dfrac{5}{12}$
11) $-21\dfrac{11}{12}$
12) -2

Exercise 3.7 Set B

1) $-\dfrac{9}{64}$
2) $-\dfrac{49}{36}$
3) $-\dfrac{1}{800}$
4) $-\dfrac{16}{81}$
5) -3
6) $-21\dfrac{7}{8}$
7) -61
8) $-\dfrac{14}{45}$
9) $-1\dfrac{5}{6}$
10) $\dfrac{5}{33}$
11) $-17\dfrac{5}{12}$
12) $-\dfrac{11}{12}$

Exercise 3.8 Set A

1) $6 + x$
2) $8 - \dfrac{1}{2}x$
3) $\dfrac{3}{4}x$
4) $\dfrac{x}{15}$
5) $\dfrac{3}{5}x - 5$
6) $11\left(\dfrac{3}{4}x\right)$
7) $\dfrac{x-3}{4}$
8) $1 + \dfrac{8}{9}x$
9) $2\left(\dfrac{1}{3} + x\right)$
10) $\dfrac{2}{7}x - \dfrac{1}{7}x$
11) $13 + \dfrac{1}{2}x = x - 3$
12) $20 + 2x = \dfrac{1}{2}x$
13) $10x - 6x - 4x = 5$
14) $\dfrac{x}{-4} + 12 = 21$
15) $\dfrac{3}{4}\left(x + \dfrac{1}{x}\right) = 33$
16) $\dfrac{1}{5}x\left(\dfrac{2}{5}x\right)\left(\dfrac{3}{5}x\right) = 50$
17) $3x - 45 = \dfrac{1}{2}x$
18) $\dfrac{2}{9}x \div 2 = 5 + x$
19) $6(14 + \dfrac{2}{8}x) = 1$
20) $\dfrac{1}{4}(5 + x) = 24$

Exercise 3.8 Set B

1) $x + (-5)$
2) $\frac{2}{3}x - 3$
3) $\frac{1}{2}x$
4) $\frac{12}{x}$
5) $6 + \frac{4}{5}x$
6) $\frac{3}{4}x\left(\frac{1}{4}x\right)(16)$
7) $\frac{x-5}{8}$
8) $\frac{2}{9}x - 1$
9) $3(x + \frac{2}{3})$
10) $\frac{2}{6}x - \frac{1}{7}x$
11) $13 - \frac{1}{4}x = 3 + \frac{1}{2}x$
12) $\frac{3x}{9} = \frac{1}{2}x$
13) $4x + 7x + 3x = 3$
14) $\frac{6}{x} - 10 = 22$
15) $\frac{1}{3}(x + \frac{1}{x}) = 12$
16) $\frac{3}{8}x\left(\frac{2}{8}x\right)\left(\frac{5}{8}x\right) = 90$
17) $2x - 72 = \frac{1}{5}x$
18) $\left(\frac{2}{9}x\right)\left(\frac{2}{3}x\right) = 5 + x$
19) $5(4 - \frac{2}{8}x) = 14$
20) $(x - 5)(3 + \frac{1}{4}x) = 29$

Exercise 4.1
1) five tenths
2) seventeen hundredths
3) thirty-nine thousandths
4) five and seven ten-thousandths
5) sixteen and thirty-five hundredths
6) four hundred twenty-six and nine tenths
7) six and one thousand two hundred thirty-six ten-thousandths
8) 0.7
9) 0.12
10) 9.003
11) 45.0006
12) 100.16
13) 356.207
14) 5023.3517
15) 2,080,000.001086
16) 2 tenths
17) 5 thousandths
18) 4 ten-thousandths
19) 6 tens
20) 7 ones
21) 3 hundredths

Exercise 4.2
1) 36.7
2) 4.7
3) 125.5
4) 90.0
5) 9.0
6) 0.1
7) 18.72
8) 5.48
9) 792.04
10) 0.90
11) 7.20
12) 1.02
13) 562.018
14) 562.0185
15) 560
16) 600
17) 46.96
18) 46.964
19) 46.9635

Exercise 4.3
1) 511.41
2) 124.633
3) 2.0
4) 41.09
5) 222.16
6) 586
7) 369.45
8) 6.47
9) 426.69
10) 1381.78
11) 596
12) 19.95
13) 147.43
14) 3.417
15) 32456.7
16) 324,567
17) 3,245,670
18) 32,456,700
19) 47.09
20) 43370
21) 45407
22) 608.5
23) 214.50671
24) 21.450671
25) 2.1450671
26) 0.21450671
27) 1.47
28) 1410.01
29) 8.888
30) 757

Exercise 4.4 Set A

1) 0.6
2) 0.5
3) 0.625
4) 0.0625
5) 1.28125
6) 1.15625
7) 0.545
8) 0.833
9) 1.857
10) 2.444
11) 23.667
12) 37.778
13) $\frac{3}{5}$
14) $\frac{2}{5}$
15) $\frac{31}{50}$
16) $\frac{37}{50}$
17) $\frac{17}{400}$
18) $\frac{19}{400}$
19) $4\frac{1}{1000}$
20) $3\frac{9}{1000}$

Exercise 4.4 Set B

1) 0.8
2) 0.75
3) 0.375
4) 0.3125
5) 1.21875
6) 1.09375
7) 0.556
8) 0.571
9) 2.231
10) 3.455
11) 44.167
12) 32.143
13) $\frac{1}{5}$
14) $\frac{3}{5}$
15) $\frac{43}{50}$
16) $\frac{49}{50}$
17) $\frac{3}{80}$
18) $\frac{11}{400}$
19) $1\frac{7}{1000}$
20) $2\frac{3}{1000}$

Exercise 4.5 Set A

1a) $224.99
1b) $80.87
2) $627.90
3) $79.39
4) $34,281
5) $1,621,710
6a) $147,781.29
6b) $2955.63
7) $16.42
8) $1.85
9) $16.49
10) $27.54
11a) $52.10
11b) −$4.24
12) $302.50
13) 24.6
14) $0.69
15) Brand X
16) $7464.83
17) $6935
18) $9231.60
19) $170.50
20) 25
21) $96
22) 19.25
23) $192.10
24) $238.50
25) 65

Exercise 4.5 Set B

1) $443.75
2) $11.25
3) $1572.29
4) $40,146
5) $3288.37
6) $2068.19
7) $17.37
8a) $3.14
8b) $40.11
9) $82.24
10) $1349.88
11) $136.89
12) $72.49
13) $86.11
14) $0.33
15) Brand B
16) 19
17) $29050.60
18) $575.00
19) $109.50
20) $37.92
21) $86.25
22) $2090.18
23) $1538.90
24) $166.67
25) $588.42
26) 15
27) Junior Store

Exercise 4.6 Set A

1) $-\dfrac{1}{8}$
2) $10\dfrac{1}{2}$
3) $4\dfrac{53}{80}$
4) $8\dfrac{24}{25}$
5) $1\dfrac{6}{10}$
6) $14\dfrac{41}{50}$
7) $-17\dfrac{1}{5}$
8) $-5\dfrac{1}{14}$
9) $-\dfrac{1}{12}$
10) $\dfrac{9}{12500}$
11) $24\dfrac{10}{13}$
12) $4\dfrac{1}{3}$
13) -80
14) $12\dfrac{1}{2}$
15) $-\dfrac{17}{40}$
16) $1\dfrac{2}{3}$
17) $-20\dfrac{31}{50}$
18) $2\dfrac{31}{100}$
19) $\dfrac{19}{20}$
20) $1\dfrac{7}{8}$

Exercise 4.6 Set B

1) $-\dfrac{1}{8}$
2) $19\dfrac{3}{5}$
3) $11\dfrac{9}{16}$
4) $12\dfrac{24}{25}$
5) $-1\dfrac{4}{15}$
6) $49\dfrac{29}{30}$
7) $-16\dfrac{2}{3}$
8) $39\dfrac{1}{10}$
9) $-1\dfrac{5}{27}$
10) $\dfrac{27}{6250}$
11) $35\dfrac{1}{6}$
12) $-2\dfrac{11}{35}$
13) $-312\dfrac{1}{2}$
14) 5
15) $2\dfrac{13}{40}$
16) $\dfrac{16}{63}$
17) $-39\dfrac{17}{25}$
18) $\dfrac{11}{100}$
19) $1\dfrac{27}{80}$
20) $-8\dfrac{3}{4}$

Exercise 5.1 Set A

1) $\dfrac{2}{3}$
2) $\dfrac{2}{1}$
3) $\dfrac{1}{3}$
4) $\dfrac{2}{3}$
5) $\dfrac{1}{4}$
6) $\dfrac{9}{11}$
7) $\dfrac{10}{7}$
8) $\dfrac{7}{10}$
9) $\dfrac{14}{45}$
10) $\dfrac{45}{79}$

Exercise 5.1 Set B

1a) $\dfrac{127}{240}$
1b) $\dfrac{113}{240}$
1c) $\dfrac{4}{113}$
1d) $\dfrac{36}{21} = \dfrac{12}{7}$
2a) $\dfrac{100}{60} = \dfrac{5}{3}$
2b) $\dfrac{180}{104} = \dfrac{45}{26}$
2c) $\dfrac{80}{284} = \dfrac{20}{71}$
3) $\dfrac{5}{27}$
4) $\dfrac{46}{85}$
5) $\dfrac{410}{649}$
6) $\dfrac{279}{100}$
7) $\dfrac{4}{15}$
8) $\dfrac{45}{19}$
9) $\dfrac{72}{19}$
10) $\dfrac{216}{205}$
11) $\dfrac{16}{1}$

12) $\dfrac{3}{10}$

13) $\dfrac{50}{73}$

14) $\dfrac{13}{10}$

15) $\dfrac{1}{1}$

16) $\dfrac{1}{6}$

17a) $0.71/lb

17b) $2.33/gallon

17c) $4.11/yard

18a) $58\dfrac{1}{3}$ mi/hr

18b) 4 mi/gal

18c) 0.5 ft/min

19) $1.75/ft

20) no

21) Buy Food Supermarket

22) $\dfrac{27}{40}$

23) $\dfrac{1}{3}$

24) $\dfrac{4}{7}$

25) $\dfrac{4}{7}$

26) $\dfrac{1}{3}$

27) $\dfrac{27}{53}$

28) $\dfrac{47}{118}$

29) $\dfrac{3}{7}$

30) $\dfrac{8}{47}$

31) $\dfrac{13}{24}$

32) $\dfrac{7}{17}$

33) $\dfrac{8}{17}$

34a) #1: $\dfrac{3701}{1900}$; #2: $\dfrac{10107}{5125}$

34b) no

34c) company #2

35a) Bookstore B

35b) $\dfrac{65}{226}$

35c) $\dfrac{2675}{4794}$

Exercise 5.2

1) yes
2) yes
3) yes
4) yes
5) 1.12
6) 1.35
7) 11.2
8) 15.75
9) 2.47
10) .30
11) 58.89
12) 65.71
13) 76
14) 18
15) 680
16) 717.95
17) 2.5 or 2 1/2
18) 0.28 or $\dfrac{5}{18}$

Exercise 5.3 Set A

1) 180
2) 27
3) 132
4a) 200
4b) $4016.60
5) $9068.50
6) $75
7) $2040
8) 600
9) 204
10) 44
11) 125.6
12a) 150
12b) 300
13a) 22.5
13b) $213
14) 3.6 cups of cream; 1.2 cups of milk
15a) 23
15b) $86.25
16) $13.68
17) 63,480
18) 6240
19) yes
20) 16
21) 105,000
22) 2000
23) 36
24) 180
25) 9375 lbs; 5625 lbs

Exercise 5.3 Set B

1) 350
2) 20
3) 48
4a) 100
4b) $3946.60
5) $6975
6) $3125
7) $425
8) $334.25
9) no
10) $66\frac{2}{3}$
11a) 300
11b) 71
12) $89.70
13) $917.50
14a) 65
14b) $129.35
15) 6.25
16) 20.83
17a) 24
17b) 1248
18) 4.5
19) $3\frac{1}{3}$
20) 18
21) 2500
22) 50
23) 70th
24) 1650
25) 180

Exercise 6.1

1) $\frac{17}{100}$
2) $\frac{23}{100}$
3) $\frac{1}{20}$
4) $\frac{3}{50}$
5) $\frac{21}{250}$
6) $\frac{31}{500}$
7) $\frac{29}{300}$
8) $\frac{47}{600}$
9) $1\frac{9}{25}$
10) $2\frac{3}{20}$
11) $\frac{7}{16}$
12) $\frac{109}{400}$
13) $2\frac{3}{8}$; 237.5%
14) $1\frac{1}{40}$; 102.5%
15) $\frac{53}{800}$; 0.06625

Exercise 6.2 Set A

1) 0.74; 74%
2) 0.12; 12%
3) $\frac{1}{50}$; 2%
4) $\frac{2}{25}$; 8%
5) $\frac{57}{1000}$; 0.057
6) $\frac{19}{1000}$; 0.019
7) 0.2; 20%
8) 0.7; 70%
9) $\frac{187}{400}$; 46.75%
10) $\frac{177}{400}$; 44.25%
11) 2.166...; 216.67% or $216\frac{2}{3}$%
12) 1.833...; 183.3% or $183\frac{1}{3}$%
13) $2\frac{3}{8}$; 237.5%
14) $1\frac{1}{40}$; 102.5%
15) $\frac{53}{800}$; 0.06625

Exercise 6.2 Set B
1) 0.26; 26%
2) 0.08; 8%
3) $\frac{1}{25}$; 4%
4) $\frac{3}{50}$; 6%
5) $\frac{61}{1000}$; 0.061
6) $\frac{43}{1000}$; 0.043
7) 0.6; 60%
8) 0.4; 40%
9) $\frac{19}{80}$; 23.75%
10) $\frac{97}{400}$; 24.25%
11) 1.66…; 166.67% or $166\frac{2}{3}$%
12) 2.33…; 233.3% or $233\frac{1}{3}$%
13) $1\frac{1}{8}$; 112.5%
14) $2\frac{3}{40}$; 207.5%
15) $\frac{43}{800}$; 0.05375

Exercise 6.3
1) 60%
2) 80%
3) 33.33 or $33\frac{1}{3}$
4) 45
5) 3
6) 3.5
7) 26%
8) 21%
9) 225
10) 187.5
11) 47.36
12) 125.13
13) 360%
14) 600%
15) 423.53
16) 385.71
17) 8.03
18) 3.17
19) 120%

Exercise 6.4 Set A
1) 5%
2) 12%
3) 40
4) 8500
5) 400
6) 201.6
7) $589.54
8) $202.46
9) $96.06
10) $427.72
11) $406.25
12) $3087.50
13) $2938.02
14) $156.58
15a) $2936.75
15b) $489.46
16) $35,452.16
17) $872.88
18) $122,850
19) 60%
20) 40%
21) $46\frac{2}{3}$%
22) 600
23a) 8700
23b) 8,625
24) 15,800
25) 2000
26) 4980
27) $70.91
28) $27\frac{3}{11}$%
29) 5%
30) 20.2%
31) 18.8%
32) 20.1%
33) yes
34) 16.4%
35) 375%
36) Judy
37a) 34.6%
37b) 13.1%
37c) no
38a) $3.10/lb
38b) 79%
38c) yes
39) $357.14
40) 16,900

Exercise 6.4 Set B

1) 87%
2) 525
3) 4675
4) 140
5) $69.93
6) $473
7) $573.92
8) $65.78
9) $4991
10) $1015
11) $303.22
12) $45,414.77
13) $1969.30
14) $74,851.80
15) 57.1%
16) $66\frac{2}{3}$%
17) $91,428.57
18a) 133,525
18b) $6,735,662.50
19) 6.7
20) 495
21) 149,970
22) 1588
23) $0.05
24) $16\frac{2}{3}$%
25) $13\frac{1}{3}$%
26) 24.1%
27) $22\frac{2}{9}$%
28) 42.7%
29) 14.3%
30) 21.7%
31) 39.2%
32) 12.6%
33a) $28
33b) 25%
33c) no
34) $428.03
35) $86,125.00

Exercise 6.5

1) $1800
2) $5760
3) $3375
4) $16,875
5) $4143.75
6) $8925
7a) $23,250
7b) $178,250
8a) $2457
8b) $47,957
9a) $400,000
9b) $475,000
9c) $875,000
10a) $15,593.25
10b) $7562.73
10c) $23,155.98
11) $1,331,085
12) $83,562.50
13a) $612,000
13b) $1,122,000
13c) $9350
14a) $3225.60
14b) $25,625.60
14c) $711.82
15a) $89,887.50
15b) $324,887.50
15c) $6016.44
16a) $1,258,943.30
16b) $10,491.19
17) $66,388.89
18) $7537.92
19a) $13,090
19b) $1576.62
20a) $70,000
20b) $4466.39
21a) $300,480
21b) $1986.51
21c) $2504
21d) We Lend Bank
22a) $3965.99
22b) $422.23
23a) $9415.66
23b) $12,711.14
23c) yes
24a) $334560
24b) $5622.47
25a) $2020.20
25b) $15.54
25c) $127.54
25d) no profit made

Exercise 7.1 Set A

1) 8
2) 6
3) 14
4) 2
5) 0
6) 0
7) 63
8) 46
9) 25
10) 51
11) $276.89
12) not applicable
13) $166\frac{2}{3}$ ft

Exercise 7.1 Set B

1) -32
2) $-80\frac{5}{9}$
3) $-2\frac{11}{20}$
4) -34
5) $-1\frac{47}{120}$
6) $2\frac{5}{8}$
7) 3
8) $1\frac{1}{2}$
9) $\frac{1}{2}$
10) -6
11) -1
12) $40\frac{1}{2}$
13) 10
14) $1\frac{3}{4}$
15) 78.8°
16) $32,250

Exercise 7.2 Set A
1) commutative property of addition
2) inverse property of addition
3) identity property of multiplication
4) commutative property of multiplication
5) distributive property
6) associative property of multiplication
7) inverse property of multiplication
8) commutative property of addition
9) inverse property of addition
10) commutative property of multiplication
11) distributive property
12) associative property of addition

Exercise 7.2 Set B
1) commutative property of multiplication
2) distributive property
3) associative property of addition
4) identity property of addition
5) identity property of multiplication
6) inverse property of multiplication
7) commutative property of multiplication
8) distributive property
9) commutative property of multiplication
10) inverse property of addition
11) commutative property of addition
12) associative property of multiplication

Exercise 7.3 Set A
1) x
2) $-a + 7b$
3) $29x^2 + 8xy - 5y^2$
4) $81m + 6$
5) $-4x + 10y$
6) $-8a + 67$
7) $15a - 23b$
8) $4x - 3y + 10z$
9) $3x + 10$
10) $-2m + 11mn - 14n$
11) $44x - 32$
12) $-15y + 18$
13) $56x - 40y$
14) $\dfrac{1}{9}x - 2$
15) $-144a + 123$
16) 8
17) 0
18) $n - 26$
19) $128a + 56b + 6$
20) $-17x + 360$

Exercise 7.3 Set B
1) $-9y$
2) $5m - 6n - 4$
3) $19x^2 + 7xy - 4y^2$
4) $76a + 10$
5) $39x - 10y$
6) $-6b + 56$
7) $20x - 14y$
8) $6x + 7y + 5z$
9) $18x + 25$
10) $18a + 17ab - 41b$
11) $13y + 32$
12) $-60x + 31$
13) $111x - 78$
14) $\dfrac{9}{8}x - \dfrac{3}{4}$
15) $75a + 4$
16) $32b - 52$
17) 0
18) $-70m - 33$
19) $190x - 180y + 9$
20) $28y - 44$

Exercise 7.4 Set A

1) yes
2) no
3) no
4) yes
5) no
6) no
7) 10
8) 56
9) 100
10) −36
11) −14
12) 5
13) 110
14) 54
15) 30
16) −100
17) $\frac{3}{5}$
18) $2\frac{1}{4}$
19) $2\frac{1}{3}$
20) $7\frac{17}{20}$
21) 2
22) −9
23) 11
24) −106
25) −12
26) $\frac{1}{25}$
27) $-2\frac{4}{5}$
28) $\frac{1}{12}$
29) $-\frac{6}{11}$
30) 1

Exercise 7.4 Set B

1) yes
2) no
3) no
4) yes
5) no
6) no
7) 56
8) 29
9) 35
10) −11
11) −39
12) 5
13) −147
14) 24
15) −350
16) 27
17) $-\frac{1}{4}$
18) $-4\frac{1}{5}$
19) $6\frac{1}{2}$
20) $4\frac{7}{9}$
21) $\frac{1}{2}$
22) −11
23) 13
24) −47
25) −4
26) $\frac{1}{225}$
27) $-1\frac{13}{27}$
28) $\frac{2}{3}$
29) $\frac{10}{21}$
30) $\frac{1}{3}$

Exercise 7.5 Set A

1) 2
2) 4
3) 22
4) −5
5) 7
6) 16
7) 2
8) $-3\frac{2}{3}$
9) −9
10) $\frac{2}{7}$
11) −5
12) $-3\frac{1}{3}$
13) −4
14) −7
15) $-3\frac{1}{2}$
16) −14
17) −1
18) $-6\frac{1}{2}$
19) $3\frac{1}{3}$
20) $-\frac{7}{12}$
21) 7

22) $-4\frac{2}{3}$

23) $2\frac{1}{4}$

24) 0

25) -3

26) 0

27) $\frac{-3}{44}$

28) 16

29) -14

30) $-2\frac{1}{10}$

31) -5

32) $\frac{7}{13}$

33) 2

34) $4\frac{1}{5}$

35) -2

36) $-16\frac{1}{2}$

37) $1\frac{5}{6}$

38) 3

39) 11

40) $-7\frac{2}{3}$

Exercise 7.5 Set B

1) 2

2) 3

3) 9

4) -3

5) 8

6) 0

7) -1

8) $-8\frac{2}{3}$

9) 18

10) $\frac{9}{16}$

11) -3

12) -5

13) 6

14) $-7\frac{5}{6}$

15) $-2\frac{1}{2}$

16) -5

17) -2

18) -4

19) -12

20) $\frac{7}{10}$

21) 5.2

22) $-6\frac{1}{3}$

23) $\frac{2}{3}$

24) 0

25) $\frac{4}{13}$

26) -16

27) $-\frac{2}{5}$

28) 10

29) -8

30) $1\frac{6}{25}$

31) -6

32) 3

33) 1

34) $-\frac{7}{15}$

35) -7

36) -28

37) $\frac{25}{16}$

38) 5

39) 12

40) $10\frac{4}{5}$

Exercise 7.6 Set A

1) $6 + x = 21$; $x = 5$

2) $8 - x = 30$; $x = -22$

3) $\frac{3}{4}x = 9$; $x = 12$

4) $\frac{x}{15} = 3$; $x = 45$

5) $2x - 5 = 13$; $x = 9$

6) $8x + 11 = -77$; $x = -11$

7) $\frac{x-3}{4} = 1$; $x = 7$

8) $1 + 3x = -23$; $x = -8$

9) $2(x + 6) = 12$; $x = 0$

10) $\frac{1}{2}x - 20 = 10$; $x = 60$

11) $13 + x = 2x - 3$; $x = 16$

12) $20 + 2x = \frac{1}{2}x$; $x = -13\frac{1}{3}$

13) $10x + 6x + 4x = -60$; $x = -3$

14) $\frac{x}{-4} + 12 = 21$; $x = -36$

15) $\frac{3}{4}(x + 8) = 33$; $x = 36$

16) $2(x - 3) = 3x + 9$; $x = -15$

17) $3x - 45 = \frac{1}{2}x$; $x = 18$

18) $10 + 4x = 5 + x$; $x = -1\frac{2}{3}$

19) $6(4 + \frac{2}{8}x) = 3$; $x = -14$

20) $7(5 + x) = 35$; $x = 0$

Exercise 7.6 Set B

1) $x + (-5) = 14$; $x = 19$
2) $3 - x = 12$; $x = -9$
3) $-10x = 100$; $x = -10$
4) $\frac{x}{12} = 2$; $x = 24$
5) $6 + 2x = -18$; $x = -12$
6) $\frac{3}{4}x - \frac{1}{4}x = 5$; $x = 10$
7) $x + 25 = 2x$; $x = 25$
8) $\frac{2}{9}x - 1 = 4$; $x = 22\frac{1}{2}$
9) $3(x + 3) = -36$; $x = -15$
10) $\frac{1}{2}x + 10 = 11$; $x = 2$
11) $13 - x = 3 + 2x$; $x = 3\frac{1}{3}$
12) $9x = x + 32$; $x = 4$
13) $4x + 7x + 3x = 56$; $x = 4$
14) $\frac{x}{6} - 10 = 22$; $x = 192$
15) $\frac{1}{3}(6x + 12) = -20$; $x = -12$
16) $4\left(\frac{3}{8}x\right) = x - 90$; $x = -180$
17) $2x - 72 = 80 + x$; $x = 152$
18) $3x + x = 5 + x$; $x = 1\frac{2}{3}$
19) $5(4 - x) = 4x + 1$; $x = 2\frac{1}{9}$
20) $3(x - 5) = 27$; $x = 14$

Exercise 8.1

1) not well-defined
2) well-defined
3) well-defined
4) well-defined
5) {31, 32, 33, 34, 35, 36, 37, 38, 39}
6) {h, a, p, y}
7) {6, 12, 18, 24, . . .}
8) {1, 3, 5, 7, 9}
9) {x | x is an even counting number between 2 and 100, inclusive}
10) {x | x is a multiple of 3}
11) {x | x is an integer}
12) {x | x is a month}
13) infinite
14) finite
15) infinite
16) infinite
17) finite
18) $\in$
19) $\in$
20) $\notin$
21) $\in$
22) $\notin$
23) $\notin$
24) $\subset, \subseteq$
25) $\subseteq$
26) $\subseteq$
27) $\not\subset$
28) $\subset, \subseteq$
29) $\subset, \subseteq$
30) $\subset, \subseteq$
31) $\subset, \subseteq$
32) true
33) false
34) true
35) true
36) true
37) false
38) false
39) false
40) true
41) false
42) true
43) false
44) b, c
45) a, c
46) b, c, d
47) a, c

Exercise 8.2

1) {1, 2, 4, 6, 8, 10}
2) {3, 4, 6, 7, 8}
3) ∅
4) U
5) {2, 4, 5, 6, 8, 9}
6) {2, 8}
7) ∅
8) {1, 3, 5, 7, 9, 10}
9) {1, 2, 3, 4, 6, 7, 8, 10}
10) {1, 3, 4, 6, 7}
11) {1, 2, 3, 4, 5, 6, 7, 8, 9, 10}
12) ∅
13) {2, 5, 8, 9, 10}
14) {2, 8}
15) {4, 6}
16) {1, 3, 5, 7, 9, 10}
17) {4, 6}
18) ∅
19) {1, 3, 7, 10}
20) {all diet cola soda pops}
21) {all caffeine-free soda pops in cans}
22) {all non-diet soda pops in cans}
23) {all caffeine diet soda pops}
24) 124
25) 19
26) 132
27) 125
28) 77
29) 58

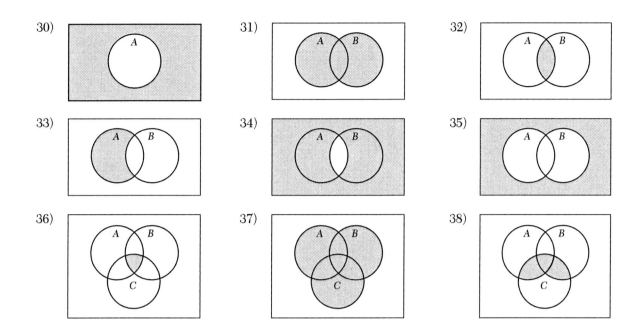

39) U = {Allison, Emily, Louis, Christopher, Roger, Staci, Mindy, Harriet, Linda, Stanley}
40) M = {Louis, Christopher, Roger, Stanley}
41) F = {Allison, Emily, Staci, Mindy, Harriet, Linda}
42) C = {Allison, Louis, Roger, Harriet, Stanley}
43) C′ = {Emily, Christopher, Staci, Mindy, Linda}
44) {Allison, Harriet}
45) {Allison, Harriet}
46) U
47) {Louis, Roger, Stanley}
48) ∅
49) {c, d, e, f, m, x, y, z}
50) {a, d, e, f, m}
51) {f, p, x, y}
52) {a, d, x}
53) {a, z}
54) {p, n}
55) {z}
56) {a, d, x, z}
57) ∅
58) {z}
59) ∅

Exercise 8.3

1) ordinal
2) cardinal
3) ordinal
4) cardinal
5) ordinal
6) cardinal
7) a, f
8) e, d
9) b
10) c
11a) no
11b) yes
11c) yes
12) 50
13) 26
14) 6
15) 5
16) 4
17) 50
18) 10
19) 4
20) 5
21) 7
22) 6
23) 5
24) 3
25) 8
26) 4
27) 3
28) 7
29) 0
30) 10

Exercise 8.4

1)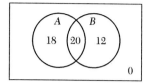

2)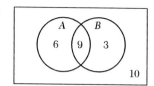

3)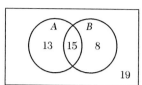

4)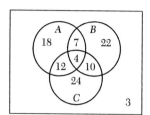

5)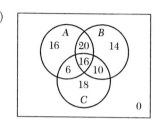

6)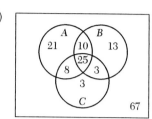

7)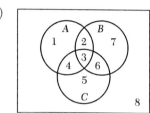

8a)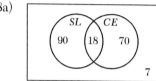

SL = Spike Lee
CE = Clint Eastwood

8b) 7
8c) 70
8d) 90

9a)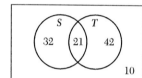

S = Soccer
T = Tennis

9b) 21
9c) 74

10a)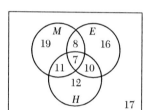

M = Math
H = History
E = English

10b) 19
10c) 16
10d) 27
10e) 17

11a)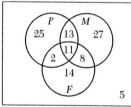

P = Pedicure
M = Manicure
F = Facial

11b) 25
11c) 16
11d) 5

12a)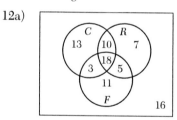

C = Cheerios
R = Rice Krispies
F = Frosted Flakes

12b) 83
12c) 18
12d) 46
12e) 27

13a)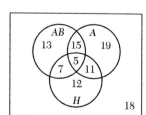

AB = Abdomin
A = Arms
H = Hips

13b) 100
13c) 13
13d) 44
13e) 42
13f) 63

Exercise 9.1

H = heads T = tails

1a) HHH HHT HTH HTT TTT TTH THT THH
1b) 8
1c) yes
1d) 1

B = boys G = girls

2a) BBB BBG BGB BGG GGG GGB GBG GBB
2b) 8
2c) yes
2d) 1

3a) (1, 1) (1, 2) (1, 3) (1, 4) (1, 5) (1, 6) (2, 1) (2, 2) (2, 3) (2, 4) (2, 5) (2, 6) (3, 1) (3, 2) (3, 3) (3, 4) (3, 5) (3, 6) (4, 1) (4, 2) (4, 3) (4, 4) (4, 5) (4, 6) (5, 1) (5, 2) (5, 3) (5, 4) (5, 5) (5, 6) (6, 1) (6, 2) (6, 3) (6, 4) (6, 5) (6, 6)
3b) 36
3c) yes
3d) 5
3e) 15
3f) 18

Let bk = black, br = brown, bl = blue, bg = beige, g = green, w = white, gr = gray; (shoes, pants, shirts)

4a) (bk, bk, bl) (bk, bk, bk) (bk, bk, g) (bk, bk, bg) (bk, bk, w) (bk, bk, gr) (bk, bg, bl) (bk, bg, bk) (bk, bg, g) (bk, bg, bg) (bk, bg, w) (bk, bg, gr) (br, bk, bl) (br, bk, bk) (br, bk, g) (br, bk, bg) (br, bk, w) (br, bk, gr) (br, bg, bl) (br, bg, bk) (br, bg, g) (br, bg, bg) (br, bg, w) (br, bg, gr)
4b) 24 outfits

Let I = Iceberg, R = Romaine, B = Boston, T = Tomatoes, C = Cucumbers, BB = Bacon Bits, CR = Croutons, IT = Italian Dressing, BC = Blue Cheese Dressing; (lettuce, toppings, dressing)

5a) (I, T, IT) (I, T, BC) (I, C, IT) (I, C, BC) (I, BB, IT) (I, BB, BC) (I, CR, IT) (I, CR, BC) (R, T, IT) (R, T, BC) (R, C, IT) (R, C, BC) (R, BB, IT) (R, BB, BC) (R, CR, IT) (R, CR, BC) (B, T, IT) (B, T, BC) (B, C, IT) (B, C, BC) (B, BB, IT) (B, BB, BC) (B, CR, IT) (B, CR, BC)
5b) 24
5c) 8
5d) 3
5e) 12

Let B = Bruschetta, FC = Fried Calamari, M = Minestrone, HS = House Salad, CS = Caesar Salad, L = Lasagna, EP = Eggplant Parmigiana, T = Tiramisu, C = Cannoli; (appetizer, soup, salad, entrée, dessert)

6a) (B, M, HS, L, T) (B, M, HS, L, C) (B, M, HS, EP, T) (B, M, HS, EP, C) (B, M, CS, L, T) (B, M, CS, L, C) (B, M, CS, EP, T) (B, M, CS, EP, C) (FC, M, HS, L, T) (FC, M, HS, L, C) (FC, M, HS, EP, T) (FC, M, HS, EP, C) (FC, M, CS, L, T) (FC, M, CS, L, C) (FC, M, CS, EP, T) (FC, M, CS, EP, C)
6b) 16
6c) 8
6d) 4

Let ML = Medieval Literature, MP = Modern Poetry, CA = College Algebra, MLA = Math for Liberal Arts, IS = Intro to Spreadsheets, WP = Web Page Design, AA = Aging in America, MA = Minorities in America, WAC = Women in American Culture; (English class, Math class, Sociology class)

7a) (ML, CA, IS, AA) (ML, CA, IS, MA) (ML, CA, IS, WAC) (ML, CA, WP, AA) (ML, CA, WP, MA) (ML, CA, WP, WAC) (ML, MLA, IS, AA) (ML, MLA, IS, MA) (ML, MLA, IS, WAC) (ML, MLA, WP, AA) (ML, MLA, WP, MA) (ML, MLA, WP, WAC) (MP, CA, IS, AA) (MP, CA, IS, MA) (MP, CA, IS, WAC) (MP, CA, WP, AA) (MP, CA, WP, MA) (MP, CA, WP, WAC) (MP, MLA, IS, AA) (MP, MLA, IS, MA) (MP, MLA, IS, WAC) (MP, MLA, WP, AA) (MP, MLA, WP, MA) (MP, MLA, WP, WAC)
7b) 24
7c) 6
7d) 4
7e) 6

Let A = Arnold, B = Bob, C = Cindy, D = Don, E = Ellen

8a) (A, B) (A, C) (A, D) (A, E) (B, A) (B, C) (B, D) (B, E) (C, A) (C, B) (C, D) (C, E) (D, A) (D, B) (D, C) (D, E) (E, A) (E, B) (E, C) (E, D)
8b) (D, A, B) (D, A, C) (D, A, E) (D, B, A) (D, B, C) (D, B, E) (D, C, A) (D, C, B) (D, C, E) (D, E, A) (D, E, B) (D, E, C)
8c) (C, A, E) (C, B, E) (C, D, E) (E, A, C) (E, B, C) (E, D, C)

Let Mr. T = Mr. Town, Mrs. R = Mrs. Rocca, Mr. C = Mr. Chang, Mr. D = Mr. Diaz, Mrs. A = Mrs. Abdu, Mrs. M = Mrs. McCann

9a) (Mr. T, Mr. D, Mr. C, Mrs. A, Mrs. M, Mrs. R)
(Mr. T, Mr. D, Mr. C, Mrs. M, Mrs. A, Mrs. R)
(Mr. T, Mrs. A, Mr. C, Mr. D, Mrs. M, Mrs. R)
(Mr. T, Mrs. A, Mr. C, Mrs. M, Mr. D, Mrs. R)
(Mr. T, Mrs. M, Mr. C, Mrs. A, Mr. D, Mrs. R)
(Mr. T, Mrs. M, Mr. C, Mr. D, Mrs. A, Mrs. R)

9b) (Mr. C, Mrs. A, Mr. D, Mr. T, Mrs. R, Mrs. M)
(Mr. C, Mrs. A, Mr. D, Mr. T, Mrs. M, Mrs. R)
(Mr. C, Mrs. A, Mr. D, Mrs. R, Mr. T, Mrs. M)
(Mr. C, Mrs. A, Mr. D, Mrs. R, Mrs. M, Mr. T)
(Mr. C, Mrs. A, Mr. D, Mrs. M, Mr. T, Mrs. A)
(Mr. C, Mrs. A, Mr. D, Mrs. M, Mrs. A, Mr. T)
(Mr. T, Mr. C, Mrs. A, Mr. D, Mrs. R, Mrs. M)
(Mr. T, Mr. C, Mrs. A, Mr. D, Mrs. M, Mrs. R)
(Mrs. R, Mr. C, Mrs. A, Mr. D, Mr. T, Mrs. M)
(Mrs. R, Mr. C, Mrs. A, Mr. D, Mrs. M, Mr. T)
(Mrs. M, Mr. C, Mrs. A, Mr. D, Mr. T, Mrs. R)
(Mrs. M, Mr. C, Mrs. A, Mr. D, Mrs. R, Mr. T)

(Mr. T, Mrs. R, Mr. C, Mrs. A, Mr. D, Mrs. M)
(Mr. T, Mrs. M, Mr. C, Mrs. A, Mr. D, Mrs. R)
(Mrs. R, Mr. T, Mr. C, Mrs. A, Mr. D, Mrs. M)
(Mrs. R, Mrs. M, Mr. C, Mrs. A, Mr. D, Mr. T)
(Mrs. M, Mr. T, Mr. C, Mrs. A, Mr. D, Mrs. R)
(Mrs. M, Mrs. R, Mr. C, Mrs. A, Mr. D, Mr. T)
(Mr. T, Mrs. R, Mrs. M, Mr. C, Mrs. A, Mr. D)
(Mr. T, Mrs. M, Mrs. R, Mr. C, Mrs. A, Mr. D)
(Mrs. R, Mr. T, Mrs. M, Mr. C, Mrs. A, Mr. D)
(Mrs. R, Mrs. M, Mr. T, Mr. C, Mrs. A, Mr. D)
(Mrs. M, Mr. T, Mrs. R, Mr. C, Mrs. A, Mr. D)
(Mrs. M, Mrs. R, Mr. T, Mr. C, Mrs. A, Mr. D)

9c) (Mr. T, Mrs. R, Mr. C, Mrs. A, Mr. D, Mrs. M)
(Mr. T, Mrs. A, Mr. C, Mrs. R, Mr. D, Mrs. M)
(Mr. T, Mrs. R, Mr. D, Mrs. A, Mr. C, Mrs. M)
(Mr. T, Mrs. A, Mr. C, Mrs. A, Mr. D, Mrs. M)
(Mr. C, Mrs. R, Mr. T, Mrs. A, Mr. D, Mrs. M)
(Mr. C, Mrs. R, Mr. D, Mrs. A, Mr. T, Mrs. M)
(Mr. C, Mrs. A, Mr. T, Mrs. R, Mr. D, Mrs. M)
(Mr. C, Mrs. A, Mr. D, Mrs. R, Mr. T, Mrs. M)
(Mr. D, Mrs. R, Mr. C, Mrs. A, Mr. T, Mrs. M)
(Mr. D, Mrs. R, Mr. T, Mrs. A, Mr. C, Mrs. M)
(Mr. D, Mrs. A, Mr. C, Mrs. R, Mr. T, Mrs. M)
(Mr. D, Mrs. A, Mr. T, Mrs. R, Mr. C, Mrs. M)

Let M^1 = husband #1, W^1 = wife #1, M^2 = husband #2, W^2 = wife #2, P = president

10a) (M^1, M^2, P, W^1, W^2) (M^1, M^2, P, W^2, W^1)
(M^2, M^1, P, W^1, W^2) (M^2, M^1, P, W^2, W^1)
(W^1, W^2, P, M^1, M^2) (W^1, W^2, P, M^2, M^1)
(W^2, W^1, P, M^1, M^2) (W^2, W^1, P, M^2, M^1)

10b) (M^1, W^1, P, M^2, W^2) (W^1, M^1, P, W^2, M^2)
(M^2, W^2, P, M^1, W^1) (W^2, M^2, P, W^1, M^1)
(M^1, W^2, P, M^2, W^1) (M^2, W^1, P, M^1, W^2)
(W^1, M^2, P, W^2, M^1) (W^2, M^1, P, W^1, M^2)

10c) (M^1, W^1, P, M^2, W^2) (W^1, M^1, P, W^2, M^2)
(M^2, W^2, P, M^1, W^1) (W^2, M^2, P, W^1, M^1)

10d) (M^1, W^1, P, M^2, W^2) (M^2, W^2, P, M^1, W^1)

Let A = Andy, B = Betty, C = Clyde, D = Dawn, E = Evan, F = Felicia

11a) (C, D, A, B, E, F) (C, D, A, B, F, E)
(C, D, A, F, E, B) (C, D, A, F, B, E)
(C, D, A, E, F, B) (C, D, A, E, B, F)
(C, D, B, A, E, F) (C, D, B, A, F, E)
(C, D, B, E, A, F) (C, D, B, E, F, A)
(C, D, B, F, E, A) (C, D, B, F, A, E)
(C, D, E, A, B, F) (C, D, E, A, F, B)
(C, D, E, B, A, F) (C, D, E, B, F, A)
(C, D, E, F, A, B) (C, D, E, F, B, A)
(C, D, F, A, B, E) (C, D, F, A, E, B)
(C, D, F, B, A, E) (C, D, F, B, E, A)
(C, D, F, E, B, A) (C, D, F, E, A, B)

11b) (B, A, C, D, E, F) (B, A, C, E, D, F)
(B, A, D, C, E, F) (B, A, D, E, C, F)
(B, A, E, D, C, F) (B, A, E, C, D, F)

11c) (E, A, C, B, D, F) (E, C, A, B, D, F)
(E, A, C, D, B, F) (E, C, A, D, B, F)
(E, A, C, B, F, D) (E, C, A, B, F, D)
(E, A, C, F, D, B) (E, C, A, F, D, B)
(E, A, C, F, B, D) (E, C, A, F, B, D)
(E, A, C, D, F, B) (E, C, A, D, F, B)

12a) 10, 11, 12, 13, 14, 20, 21, 22, 23, 24, 30, 31, 32, 33, 34, 40, 41, 42, 43, 44

12b) 10, 12, 13, 14, 20, 21, 23, 24, 30, 31, 32, 34, 40, 41, 42, 43

12c) 10, 12, 14, 20, 22, 24, 30, 32, 34, 40, 42, 44

12d) 10, 20, 30, 40

13a) 200, 202, 204, 207, 209, 220, 222, 224, 227, 229, 240, 242, 244, 247, 249, 270, 272, 274, 277, 279, 290, 292, 294, 297, 299, 400, 402, 404, 407, 409, 420, 422, 424, 427, 429, 440, 442, 444, 447, 449, 470, 472, 474, 477, 479, 490, 492, 494, 497, 499, 700, 702, 704, 707, 709, 720, 722, 724, 727, 729, 740, 742, 744, 747, 749, 770, 772, 774, 777, 779, 790, 792, 794, 797, 799, 900, 902, 904, 907, 909, 920, 922, 924, 927, 929, 940, 942, 944, 947, 949, 970, 972, 974, 977, 979, 990, 992, 994, 997, 999

13b) 204, 207, 209, 240, 247, 249, 270, 274, 279, 290, 294, 297, 402, 407, 409, 420, 427, 429, 470, 472, 479, 490, 492, 497, 702, 704, 709, 720, 724, 729, 740, 742, 749, 790, 792, 794, 902, 904, 907, 920, 924, 927, 940, 942, 947, 970, 972, 974

13c) 200, 204, 220, 240, 244, 272, 292, 400, 404, 420, 440, 444, 472, 492, 704, 720, 740, 744, 772, 792, 900, 904, 920, 940, 944, 972, 992

13d) 207, 209, 247, 249, 279, 297, 407, 409, 427, 429, 479, 497, 709, 729, 749, 907, 927, 947

14a) amex, amxe, aemx, aexm, axem, axme, maex, maxe, meax, mexa, mxae, mxea, eamx, eaxm, exam, exma, emax, emxa, xame, xaem, xema, xeam, xmae, xmea

14b) amxe, axme, emxa, exma,

14c) mxea, mxae, xmea, xmae, amxe, axme, emxa, exma, eamx, eaxm, aemx, aexm

15a) ttt, tto, ttp, tot, too, top, tpt, tpo, tpp, ott, oto, otp, oot, ooo, oop, opt, opo, opp, ptt, pto, ptp, pot, poo, pop, ppt, ppo, ppp

15b) top, tpo, otp, opt, pto, pot

15c) ttt, ttp, tot, top, tpt, tpp, ott, otp, oot, oop, opt, opp, ptt, ptp, pot, pop, ppt, ppp

15d) top, pot

16a) em38, em83, e3m8, e8m3, e38m, e83m, me38, me83, m3e8, m8e3, m38e, m83e, 3em8, 3me8, 3e8m, 3m8e, 38em, 38me, 8em3, 8me3, 8m3e, 8e3m, 83em, 83me

16b) em38, em83, me38, me83
16c) e3m8, e8m3, m3e8, m8e3
16d) 8m3e, 83me

Exercise 9.2

1a) BBB BBG BGB BGG GGG GGB GBG GBB
1b) 8
2) $2^6 = 64$
3) 96
4) $2^{12} = 4096$
5a) 3125
5b) 120
5c) 625
5d) 96
6a) 90
6b) 81
6c) 9
6d) 45
6e) 16
7a) 900
7b) 648
7c) 100
7d) 450
7e) 8
8a) 9000
8b) 4536
8c) 1344
8d) 2520
8e) 112
9a) 2401
9b) 840
9c) 360
9d) 686
10a) 12
10b) 24
10c) 6
10d) 6
11) 67,600
12a) 5040
12b) 240
12c) 1440
13a) 3,628,800
13b) 28,800
13c) 240
13d) 120
13e) 28800
14a) 100,000
14b) 30,240
14c) 10,000
15) 3125
16) 32
17) 5,475,600
18a) 5040
18b) 576
18c) 144
19a) 720
19b) 120
19c) 240
19d) 72
20a) 12,966,811,200
20b) 5,085,024
20c) 530,122,320
21) 175,219,200,000
22a) 120
22b) 40,320
22c) 720
22d) 2880
23a) 1000
23b) 3,628,800
23c) 13,168,189,440,000
24) 6,400,000
25) 180
26a) 39,916,800
26b) 840
26c) 172,800

Exercise 9.3

1) 35%
2) 12%
3) $\frac{2}{5}$
4) 0.98
5) yes
6) yes
7) no
8) yes
9) no
10) yes
11) $\frac{1}{2}$
12) $\frac{4}{5}$
13) $\frac{19}{42}$
14a) yes
14b) $\frac{1}{4}$
14c) yes
14d) $\frac{1}{2}$
15) $\frac{1}{4}$
16b) 1
16c) $\frac{1}{16}$
16d) 4
16e) $\frac{1}{4}$
16f) $\frac{15}{16}$
17b) 4
17c) $\frac{1}{4}$
17d) 1
17e) $\frac{1}{16}$
17f) $\frac{5}{16}$
17g) $\frac{1}{16}$
18b) $\frac{5}{36}$
18c) 0
18d) $\frac{1}{12}$

18e) $\dfrac{5}{12}$

18f) $\dfrac{1}{6}$

18g) $\dfrac{1}{2}$

18h) $\dfrac{11}{12}$

19a) $\dfrac{1}{13}$

19b) $\dfrac{1}{13}$

19c) $\dfrac{1}{2}$

19d) $\dfrac{1}{4}$

19e) $\dfrac{1}{52}$

19f) $\dfrac{1}{26}$

19g) $\dfrac{3}{13}$

19h) $\dfrac{4}{13}$

19i) $\dfrac{5}{13}$

19j) $\dfrac{12}{13}$

20a) $\dfrac{1}{9}$

20b) $\dfrac{1}{3}$

20c) $\dfrac{8}{9}$

20d) $\dfrac{2}{9}$

20e) $\dfrac{4}{9}$

20f) $\dfrac{7}{9}$

21a) $\dfrac{2}{9}$

21b) $\dfrac{1}{3}$

21c) $\dfrac{5}{18}$

21d) $\dfrac{2}{9}$

21e) $\dfrac{1}{2}$

21f) $\dfrac{7}{12}$

21g) $\dfrac{8}{9}$

21h) $\dfrac{8}{9}$

22a) $\dfrac{2}{13}$

22b) $\dfrac{4}{13}$

22c) $\dfrac{1}{2}$

22d) $\dfrac{1}{26}$

22e) $\dfrac{3}{13}$

22f) $\dfrac{7}{13}$

22g) $\dfrac{4}{13}$

22h) $\dfrac{5}{13}$

22i) $\dfrac{5}{52}$

22j) $\dfrac{4}{13}$

22k) $\dfrac{7}{13}$

22l) 1

22m) $\dfrac{10}{13}$

23a) $\dfrac{119}{398}$

23b) $\dfrac{51}{398}$

23c) $\dfrac{235}{398}$

23d) 0

23e) $\dfrac{15}{398}$

23f) $\dfrac{263}{398}$

23g) $\dfrac{263}{398}$

23h) $\dfrac{209}{298}$